T5-BYG-608
WITHDRAWN

Biotechnology Policy across National Boundaries

Biotechnology Policy across National Boundaries

The Science-Industrial Complex

Darrell M. West

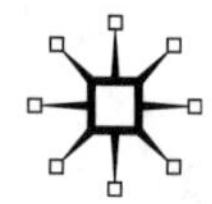

BIOTECHNOLOGY POLICY ACROSS NATIONAL BOUNDARIES

First published in 2007 by
PALGRAVE MACMILLAN™
175 Fifth Avenue, New York, N.Y. 10010 and
Houndmills, Basingstoke, Hampshire, England RG21 6XS.
Companies and representatives throughout the world.

PALGRAVE MACMILLAN is the global academic imprint of the Palgrave Macmillan division of St. Martin's Press, LLC and of Palgrave Macmillan Ltd. Macmillan® is a registered trademark in the United States, United Kingdom and other countries. Palgrave is a registered trademark in the European Union and other countries.

ISBN-13: 978-1-4039-7251-4
ISBN-10: 1-4039-7251-6

Library of Congress Cataloging-in-Publication Data is available from the Library of Congress.

A catalogue record for this book is available from the British Library.

Design by Macmillan India Ltd.

First edition: August 2007

10 9 8 7 6 5 4 3 2 1

Printed in the United States of America.

To Karin, who knows how to cross national borders

Table of Contents

Preface		ix
Chapter 1	The Globalization of Innovation	1
Chapter 2	Science-Industry Collaboration	15
Chapter 3	In Vitro Fertilization and Assisted Reproductive Technology	31
Chapter 4	Genetically Modified Food	49
Chapter 5	Cloning	69
Chapter 6	Stem Cell Research	85
Chapter 7	Chimeras	101
Chapter 8	Pharmaceutical Companies, Biotechnology, and Health Care	115
Chapter 9	Whose Ethical Standards?	131
Notes		145
Bibliography		167
Index		187

Preface

Scientists now have the ability to produce genetically modified foods and organisms. They can clone living tissue and have discovered ways to use stem cells to regenerate tissue and fight human diseases. In vitro fertilization allows people to overcome fertility problems. Cells from humans and rabbits can be used to create new, cross-species organisms known as chimeras. Yet despite the prevalence of these advances, little is understood about the role of the state in promulgating biotechnology policy. Who decides what biotechnologies to encourage? How are new initiatives funded and regulated? What role do large multinational pharmaceutical companies play? Whose ethical standards are used to judge these plans?

This book seeks to increase our understanding of biotechnology policy by analyzing how new advances are financed, regulated, and spread. I argue that a "science-industrial complex" based on universities, businesses, and nongovernment organizations has emerged that fuels biotechnological innovation. Much of this innovation is global in nature and independent of state control. As scientists collaborate across national borders, a new kind of globalization has emerged that is quite different from the traditional, economy-based globalization.

Economic globalization centers primarily on financial transactions and liberalization of cross-border trade. Many of the debates surrounding this type of globalization are concerned with how global trade affects the material conditions of nations, industries, and workers. Biotech globalization, by contrast, involves plant, animal, and human regeneration and the effect of boundary lowering on the sanctity of life. This new form of globalization is controversial because it raises ethical dilemmas related to basic human values.

In this research, I analyze the role of the public, private, and nonprofit sectors in promulgating new biotechnologies. I examine five cases (stem cell research, cloning, chimeras, genetically modified food, and in vitro fertilization) in a number of different countries (mainly the United States, Great Britain, France, Germany, India, China, Korea, and Japan) to see what role the state plays in promoting biotechnology. Through an analysis of ethics, finance, regulation, and decision making, I study biotech globalization as a cross-national process.

I focus on these particular cases because they are at the forefront of controversies over biotechnology. Ever since Louise Brown was created as the world's first test-tube baby in 1978, in vitro fertilization has become a common technique for overcoming fertility problems in many countries. The production of Dolly as the first cloned sheep in 1997 placed that topic squarely in the middle of public discussion. The use of genetically modified foods and chimeras over the past decade has sparked very different reactions in various countries. And most recently, stem cell research has emerged as a hot-button issue for scientists, business people, and public officials around the world.

Of the five cases, in vitro fertilization is the one that has generated the least controversy and the most limited state oversight. Genetically modified foods, in contrast, have been the object of intense oversight in some countries but not others. Cloning (especially that involving humans) has been regulated in most, but not all, nations. Stem cell research is controversial nearly everywhere, is the object of considerable public debate, and faces stringent state oversight around the world. Chimeras have attracted surprisingly little public or government attention despite the potentially far-reaching nature of the research.

The focus of scholarly research on biotechnology has thus far been too narrow. Most projects have been limited to individual country studies or comparative studies of small scope that focus on Western players and virtually ignore Far Eastern nations. The former approach lacks the virtue of comparative study. The advantage of examining more than one country is that it allows one to see variation and understand why different places react to the same technology in distinctive ways. The latter approach, a comparative study limited to nations such as the United States, Great Britain, Germany, or France, meanwhile is incomplete. It ignores the significance of biotech innovation in Asia. Since many Western countries have placed more restrictions on biotech than some Asian nations, it behooves scholars to investigate places such as China, India, and Korea to see how their approach to biotech contrasts with that of Western nations. In some respects, these Asian countries are following a different path and it is crucial to understand how they are handling these issues.

In hopes of better understanding biotech decision making, I adopt a cross-national approach to see how different countries deal with biotechnology. A broad comparison allows researchers to look at decision-making structures, group demands, industry composition, religion, and political culture to see why different countries have developed varying approaches to biotech policy. In addition, focusing on a variety of nations helps to see larger patterns in biotechnology such as what I term "country-shopping" by scientists and businesses, and how biotech innovators play countries off against each other in order to gain autonomy for their research.

The globalization of innovation has produced the most massive spurt in biotechnology in world history, yet it raises a host of questions regarding its long-term significance. If cross-national partnerships have challenged the primacy of the state in setting biotechnology policy and if innovation has undermined the ability of governments to regulate scientific activities, have we created a policy area that is beyond the control of political leaders? If so, this poses a number of interesting social, political, and ethical issues for scientists, politicians, and the general public.

In chapter 1, I discuss how science and technology have become a global enterprise and the role of the state in shaping biotechnology policy. Universities and corporations collaborate across national boundaries and technology transfers are commonplace. The result has been dramatic increases in scientific publications and patents and the diffusion of new technology. I look at the global infrastructure of this "science-industrial complex," and the risks this raises for globalization. Biotech globalization poses a number of problems for the international system, and it is important to understand how it differs from the economic globalization that has dominated the discussion to date.

In chapter 2, I examine the collaboration between science and private business on biotech research. Close collaboration between these sectors is common in many countries. There are a number of features that have brought industry and science close together into a science-industrial complex. Deregulation in the public sector has weakened state capacity to oversee private interests. Corporate partnerships with higher education have made scientists more dependent on industry financing for research support. The emergence of country-shopping and scientist-buying allows researchers and businesses to play nations off against each other and thereby gain considerable autonomy for themselves. Finally, the prevalence of contraband technology in global trade means that even when governments seek to regulate the science-industrial complex, such efforts are not always very successful.

Chapter 3 presents a case study of in vitro fertilization. Fertility treatments and test-tube pregnancies represent some of the earliest applications in the contemporary period. Reproductive technologies are an area that has seen the least public sector regulation in the biotech sphere. Many approaches to in vitro fertilization remain private decisions between medical professionals and patients. Most governments around the world have few restrictions in place and the result has been a rapid proliferation and acceptance of this life-generating technology.

In chapter 4, I take a look at the biotechnology of genetically modified foods that has attracted stringent public sector regulation in some countries, but not others. For example, the United States has adopted a laissez-faire approach to genetic modifications that has defined new foods as valuable commodities for commerce and trade. Agribusinesses have invested large amounts of money in research and lobbied public officials for nonregulation. In contrast, a number of European and Asian nations have expressed greater alarm over this area and have enacted strict regulations concerning research on and use of genetically modified foods. The cross-national comparison of this area shows how political, economic, and social factors explain the different regulatory reactions on this topic.

Chapter 5 presents a case study of a cloning biotechnology that has seen more serious oversight by most countries. Governments have devoted serious effort to restrict and regulate research on human cloning and even some types of cloning involving other organisms. Concerns raised by religious authorities or ethicists have led to controls on what can and cannot be undertaken. However, this effort has not been entirely successful. Scientists have migrated to locales with fewer restrictions in order to undertake research forbidden in their own homelands. These responses have limited the ability of the state to regulate this biotechnology.

In chapter 6, I investigate stem cell research, which is the biotech field with the strongest and most consistent regulation around the world. Because stem cell research has been defined as life altering, it has generated the most controversy. Many governments have placed sharp restrictions or outright bans on work in this area. This is especially the case in regard to embryonic stem cell research. For this endeavor, biotechnology has become extremely contentious and has moved from a relatively invisible concern of experts to a public issue that engages citizens and groups at large. The expansion of controversy to the public sphere has made this the most contentious biotech of all.

Chapter 7 looks at chimeras, which are genetic hybrids that are created across two or more species. Most cross-species fusions are not controversial. Indeed, crossbreeding is a centuries-old practice, both in terms of plants and animals. However, when new technologies yield organisms with the cells of human beings, the ethical challenges increase dramatically. Ethicists ask what proportion of an organism's genetic material has to be human for the organism to be considered a person? Despite the fundamental nature of this question, most governments have few rules regulating research in this area. This is problematic given the possible long-term dangers of this biotechnology.

Chapter 8 investigates the role of large multinational pharmaceutical companies in biotechnology. I review the funding of biotech, industry policies to facilitate innovation, patent and intellectual property issues in biotechnology, and ramifications for consumers and patients in the health care area. Basically, I argue that biotech offers considerable advantages to the science-industrial complex, but poses some risks to consumers.

Chapter 9 examines ethical controversies associated with biotech globalization. I contrast nations governed by "international political economy," which tend to adopt permissive biotech policies, with those influenced by "religious political economy," whose policies generally are more restrictive. I discuss whose ethics should play a role in biotechnology decisions and who should act as decision-makers: companies, scientists, universities, public regulators, elected officials, international bodies, or the general public?

I am grateful to many people for their help on this project. Emily Dietsch provided the most wonderful research assistance . She found articles dealing with biotechnology and located obscure information for me. She also edited the entire manuscript and provided helpful comments that clarified the book's argument. In addition, Tyler Rowley located articles and research studies that were very helpful in thinking about biotechnology. Funding for this project was provided by the Brown University Taubman Center for Public Policy. Anthony Wahl, senior editor at Palgrave Macmillan, deserves a thank-you for his insightful commentary on this book. He made a number of suggestions that improved the presentation of the research. Macmillan India Ltd. did a great job overseeing the copy editing of the manuscript. I am very grateful for the help of all these individuals and organizations.

Chapter 1

The Globalization of Innovation

On St. Kitts island in the Caribbean, a lab operated by a Yale University professor transplants human stem cells into live monkeys. This is a practice that is restricted in some societies as an inhumane and dangerous experiment. When the United States prohibited some forms of cloning, scientists in Great Britain rushed to produce Dolly as the world's first cloned sheep. Chinese officials were shocked recently when farmers in their country planted genetically modified rice contrary to national laws. Techniques based on in vitro fertilization have become common around the world, regardless of the types of political or religious systems.

In areas from genetically modified foods and in vitro fertilization to cloning, stem cell research, and chimeras, innovation increasingly has moved beyond national boundaries. Technology transfers are common and it is difficult for governments to regulate diffusion. Contraband technology prohibited from sale somehow evades government rules on cross-border movement. Scientists engage in what I call "country-shopping" to find locales with the most favorable rules for research. New social networks consisting of academics, members of nongovernmental organizations, and business representatives push the envelope of innovation in ways not imagined in past decades.

For observers of the political process, these kinds of technology transfers raise the important question of whether the state is obsolete in the regulation of biotechnical innovations. Has the traditional role of the public sector been undermined by the rise of technical experts, the emergence of country-shopping by scientists seeking freedom from regulation, the growth of nongovernmental organizations, and the power of multinational pharmaceutical corporations? Although states retain a significant role in funding and regulating new technology, the ability of governments to oversee innovation has been sharply circumscribed.

In this book, I argue that a "science-industrial complex" has emerged in the biotechnology area that has dramatically weakened the role of government. While this is not a uniform development in every nation, an alliance of corporations, universities, and nongovernmental organizations in many places has gained power at the expense of the public sector. This shift in the relative influence of state and non-state actors has led to the virtual deregulation of many biotechnologies and the liberation of innovation from geopolitical constraints. Using case studies of biotechnology areas, I show that despite public and leadership concern about the dangers of these innovations, many new initiatives operate beyond the control of the public sphere.

The Science and Technology Revolution

There has been a dramatic increase in technological innovation around the world. The number of patents filed over the last two decades has risen significantly. In 1985, for example, around 75,000 patents were filed in the United States and 41,000 in Europe. By 2000, these numbers had jumped to 180,000 for the United States (up 140 percent) and 110,000 in Europe (up 168 percent).[1]

Biotechnology and information technology are among the areas showing the greatest growth. Advances in genomics and biology have fueled the invention of new diagnostic tools. It now is possible to use genetic testing to check for health risks to particular illnesses. The hope is that someday, doctors will be able to use DNA science to target treatments more effectively. For example, if it is known that people of a particular genetic makeup are likely to suffer serious side effects from certain cancer therapies, doctors will avoid those therapies, thereby sparing patients unnecessary treatment complications.

The number of scientific publications worldwide has increased from 1981 to 2003. According to *Science Watch*, a commercial newsletter that monitors trends in basic research, there were around 440,000 publications in 1981, but this number skyrocketed 82 percent to 800,000 in 2003. This reflects the premium many countries are placing on technical innovation and the need to communicate results through professional networks.

There is variation from country to country in the number of scientific publications. In looking at breakdowns of science-oriented publications, the United States leads with just under 40 percent of publications. Great Britain, Germany, and Japan each have around 8 percent of the world's publications, France and China have 5 percent each, Canada and Italy have 4 percent each, and the rest are scattered among other nations. However, China's percentage has risen dramatically over the past two decades, going from near zero in 1981 to 5 percent in 2003.[2]

International collaboration on research has increased. Around 45 percent of the world's scientific publications have at least one American researcher, up from previous years. In the same vein, around 25 percent of U.S. publications include non-American authors, a figure that is considerably higher than the 10 percent of publications having an international collaborator in the late 1980s.[3]

Global innovation has risen because many countries around the world see the burgeoning of technology as a boon to trade and national development. Technology is a valuable export commodity in many lands. Trade now constitutes nearly 30 percent of global Gross Domestic Product (GDP). In 1970, it represented little more than 10 percent of global GDP.[4] Countries trade minerals, raw materials, information, and technical services.

Technology is seen as key to national development and mass communication. Nations that foster innovation in biotechnology, nanotechnology, computers, and engineering experience more economic growth.[5] The expansion of broadband Internet access and cell phone usage fosters improved communications, enhances economic growth, and leads to better national and international integration.

In a high-tech era, there is a clear association between new technologies and national innovation. Developed countries put considerable resources into research in hopes not only of strengthening military capabilities, but also of seeing commercial spin-offs in the private sector. New inventions have arisen from the space program, satellite technology, weapons development, and military procurement. The space program has sent people into outer space and has fostered a wide range of technical applications for society as a whole. Among other inventions, it has facilitated the creation of Teflon, satellite communications, longer-lasting batteries, microwave technology, and wireless telephony.[6]

Through greater understanding of basic science, experts are applying insights from a variety of fields to areas such as health care, education, transportation, defense, and communications. The Internet and e-mail speed dissemination of knowledge around the globe, with the result fueling a science and technology revolution. Innovation is globalized, and scientists are able to learn about the inventions of one another.

The Rise of a Science-Industrial Complex

Technology has fueled the rise of a "science-industrial complex," similar to what used to be called the "military-industrial complex." Private companies, universities, and nonprofit organizations collaborate across national boundaries and export innovation from country to country. As shown in chapter 2, there is close collaboration between scientists and businesses. There is big money to be made in technical spin-offs, so a variety of organizations have clear incentives to produce new products.

The United States spends 2.8 percent of its GDP on research and development. This is less than the 4.3 percent spent by Sweden, 3.1 percent by Japan, and 3.0 percent by South Korea, but higher than that of Germany (2.5 percent), France (2.2 percent), Canada (1.9 percent), or Great Britain (1.9 percent). Europe as a whole devotes 1.9 percent to research and development, while industrialized nations spend around 2.3 percent.[7]

If one adds together all the science and technology workers in the United States as a percentage of the workplace, 33 percent of American employees have science or technology positions. This is slightly less than the 34 percent figure for the Netherlands and Germany, but higher than the 28 percent in France and Canada, respectively.[8]

Technology is a big business in industrialized countries. The global growth rate for high-tech products has increased by 6.5 percent in recent years, far higher than the 2.4 percent growth for other kinds of manufactured items. High-tech products now constitute about

23 percent of manufacturing output in the U.S., 31 percent in South Korea, and 18 percent in France and Great Britain, respectively.[9]

The productivity in this area has fueled considerable demand for those with science and engineering expertise. Thirty-eight percent of Korean students now earn degrees in science and engineering, compared to 33 percent for Germany, 28 percent for France, 27 percent for Great Britain, and 26 percent for Japan. The United States has fallen behind in this area. Despite great demand for this kind of training, only 16 percent of American graduates have backgrounds in science and engineering.[10]

In America, the private sector surpassed the federal government in 1980 in terms of the amount of money spent on research and development. By 2003, commercial companies were providing 68 percent of the $283 billion spent on research and development, compared to 27 percent from the federal government and the rest from other sources. Of this total, $113 billion comes from the federal government, while $170 billion comes from the private sector. According to information from the National Science Board, the percentage of research and development spending coming from the federal government has dropped from around 63 percent in the early 1960s to 27 percent today, while that of the private sector increased from 30 to 68 percent.[11]

A similar pattern is found in many industrialized nations. The commercial sector in most countries provides a far higher share of research and development funding than does the government. For example, the private sector in Japan provides 72 percent of funding. In Germany, commercial companies provide 66 percent, while in France, private enterprise provides 52 percent of the research funding. In most of these places, relatively little investment comes from the public sector.[12]

The extensive reliance by universities on commercial investment has created considerable concern about the science-industrial complex. For example, former Harvard University president Derek Bok criticizes what he calls the "commercialization of higher education."[13] As universities have looked to the private sector for sources of new revenue, scientists have lost their traditional independence. Rather than searching for new knowledge, researchers are undertaking projects at the behest of corporations and slanting their research towards industry goals and priorities.

In her book *University, Inc.: The Corporate Corruption of Higher Education,* journalist Jennifer Washburn takes the argument one step further. She argues that rather than merely commercializing higher education, the close economic ties between science and industry have corrupted universities and raised fundamental conflicts of interest that taint the scientific enterprise.[14] Increasingly, professors are acting as outside consultants for private companies and engaging in research designed to serve corporate interests. The result has been the degradation of academic work and the undermining of higher education into what she calls the "academic-industrial complex."

Part of the close ties between industry and higher education arose from university needs for new revenue sources. The cost of new biotech labs and global science has led a number of schools to seek financing from the private sector. Federal legislation has encouraged universities to move in this direction. In 1980, Congress passed the Bayh-Dole Act, which gave university professors funded by federal grants the right to seek patents for their scientific products.[15] Although the idea was to make sure that scientific knowledge was widely disseminated, the legislation provided financial incentives for universities to commercialize basic knowledge. Over the following 25 years, this legislation succeeded beyond its wildest hopes. From a sparse 95 patents to universities in 1965, the number rose to 3,200 university patents in 2000.[16]

In his book *Science in the Private Interest,* ethics professor Sheldon Krimsky describes the "gold rush mentality" that developed in the late twentieth century in higher education.[17] Universities created intellectual property offices, patented new products, and encouraged faculty start-ups that would bring in large amounts of money. Rather than encourage knowledge for its own sake, scientists saw virtues in commercializing their expertise. Not only did they make a lot of money, they gained new clout for revenue generation within universities themselves.

By 2004, American universities had collected over $1 billion resulting from licenses, patents, and private start-up companies. In that year alone, 425 commercial firms were established, up from 348 in 2003, 364 in 2002, and 402 in 2001. In addition, these institutions of higher education filed 8,185 patent applications in 2004.[18]

This commercialization of scientific inventions is not limited to the United States. Increasingly, European and Asian universities are creating commercial spin-offs and filing patents on their new applications. Similar to the U.S. Bayh-Dole legislation enacted in 1980, many non-American countries have changed their patent laws to give ownership rights over new creations to universities and academics. Professors can share in ownership and profits, but generally this is in conjunction with the institute where the research takes place. Of European nations, only Italy and Sweden have retained the so-called professor's privilege of sole ownership on patents and inventions. The European Union now provides $4 billion to commercialize academic research. Individual nations such as Great Britain, France, Germany, and the Netherlands also provide financial support of transfer technology to the private sector.[19]

Is the State Obsolete in Biotechnology Policy?

The decline in financial support that many national governments provide for research and development and the close collaboration between science and industry raise the question of whether the state has become obsolete in managing biotech policy. Of the major policy areas, such as health, education, and welfare, biotechnology is the subject where the state exercises its weakest control. The very nature of technology itself makes it difficult for the government to control. Much of research and development is decentralized and funded outside of government. Inventions are made by small teams of researchers, often collaborating across national boundaries. Transfer of knowledge is difficult to regulate because scientific exchanges take place through informal networks, such as personal conversation, e-mails, or scientific meetings.

The technical nature of the subject means that experts, policy communities, and social networks dominate communications and innovation. Much as the state finds it difficult to regulate artistic expression or cultural flows, it is nearly impossible to control technical experts. They meet informally, exchange ideas, and publish innovative ideas. A considerable amount of work in this area is abstract knowledge or technical applications. These qualities make it very difficult for states to restrict information flows.

For these reasons, biotechnology represents an area that has been liberated from geography. Transaction and communications costs are low, and innovation is quickly transferable between individuals or organizations. Since knowledge can be diffused through scientific publications, face-to-face conversations, or electronic communications, it is difficult for governments to control the pace, diffusion, or dissemination of innovation. Nonstate actors such as universities, corporations, and nonprofit organizations have risen in power in terms of knowledge creation and dissemination, which further challenges the ability of the state to restrict access across countries.

This argument notwithstanding, some researchers have suggested that the state remains potent. For example, political scientist Daniel Drezner has written that the state remains a significant force for institutional control in the area of technology. Using a case study of global governance of the Internet, he argues that state power operates through a variety of mechanisms, such as coercion, inducements, delegation, third-party organizations, and outside groups. States do not have to control a policy venue directly for government influence to be felt, he says.[20] They can use subtle and indirect mechanisms to exercise power.

In his book *Governing the Market,* political economist Robert Wade argues that states hold considerable influence over economic development. Relying on analysis of East Asian countries, he finds that newly industrialized countries require strong states and that they need sympathetic political institutions to maximize their growth potential.[21] In a related vein, scholar Atul Kohli writes that state-level development (the so-called developmental state) is an effective strategy for promoting economic growth.[22]

But other researchers dispute these interpretations. Political scientist Peter Haas argues that it is difficult for governments to regulate technical fields because experts form policy communities and thereby exercise independent power. These "epistemic communities" represent a source of coordination, communication, and decision making. When experts agree on a matter, governments tend to follow their viewpoint. This suggests that it is not government that is dictating public policy, but private networks of technical experts.[23]

Planner AnnaLee Saxenian makes a related point about social networking in regard to computer innovation. Looking at the emergence of Silicon Valley and Route 128 around Boston, she argues that the social networking of universities, computer entrepreneurs, and military contractors was the key factor in fostering a culture of innovation on microcomputers. Entrepreneurs had regular contact with university professors, and this created a culture sympathetic to new computer products.[24]

Another force for the liberation of civil society from government is the growth of nongovernmental organizations. Professor Paul Wapner looks at environmental activism around the world and demonstrates that environmentalists have altered state decision making and become a major political force through their activism. The title of his article, "Politics Beyond the State," summarizes his major theme, that there now exists a global civil society that transcends national boundaries.[25]

Still others have claimed that the power of the state has been eroded by the power of multinational corporations. For example, scholar Virginia Hauffler examines the "public role" of private corporations and finds companies have become adept at employing economic clout to extend their influence over government.[26] Using case studies of high-tech enterprises in four Chinese communities, political observer Adam Segal argues that nongovernmental firms were the key ingredients in China's technical leap forward.[27] Scholar Richard Doner looks at statism in five Asian countries and concludes that state strength in these nations is much more limited than commonly thought.[28] And, in an analysis of developed countries, political economist John Zysman argues that "financial markets are one element that delimits the ways in which business and the state interact." Because they decide "who gets what," he says markets are a form of political and social control independent of the government.[29]

Those who defend the significance of the role of the state have developed concepts that blend state power with social context. For example, sociologist Peter Evans discusses the notion of "embedded autonomy" as a way of summarizing his argument that the state matters. He notes that state power is most effective when it is embedded in informal social structures that provide organization and coherence.[30] In a related vein, analyst Jon Pevehouse suggests that governments use nongovernmental organizations (NGOs) to influence the policies proposed and implemented by outside groups. Rather than being autonomous institutions, NGOs link into state authority, making it difficult to separate the public sector from civil society.[31]

Similarly, policy analyst Sheila Jasanoff demonstrates that the combination of political, institutional, and cultural forces leads some countries, especially in Europe, to regulate biotechnology quite effectively. Using a comparative study of the United States, Great Britain, and Germany as well as the European Union, she shows that on many biotech topics, Germany has regulated innovation more strictly than has the United States or Great Britain. Germany's history and political culture places a greater premium on regulating uncertain risks than is true for the other places. Its historical and cultural background gives it moral authority to regulate domains that other countries are not able to control.[32]

These blended approaches offer the advantage of conceptual subtlety in attempting to capture the complex ways in which public, private, and nonprofit sectors interact. Their strong point is the ability to document subtle and indirect forms of power (sometimes hidden from public view) and show how the government exercises power behind the scenes. For example, Jasanoff demonstrates that cultural forces can affect institutional design, and thereby perpetuate this power in the policymaking process.

But this approach suffers because it lumps sectors together in a way that makes it virtually impossible to disentangle them. If the public sector is embedded within social structures, political culture, the private sector, and nongovernmental organizations, it is difficult to answer questions about the relative power of each sector. Embedded autonomy draws attention to sector mutualism, but does not resolve the fundamental question of who shapes various policy areas. This limits analysts' ability to answer basic questions about political power.

This inability is particularly problematic in the case of biotechnology policy because the growth of nongovernmental organizations, the power of multinationals, the rise of expert policy communities, and country-shopping by scientists looking for lax rules and enforcement suggests that the globalization of innovation has undermined state control.[33] The large amount of research and development provided by companies and universities weakens the power of government. Similarly, the emergence of powerful non-state actors cedes significant public authority to private individuals and organizations.[34]

New versus Old Globalization

The new globalization that has emerged is very different from the old variety about which many researchers wrote. Much of the early literature on globalization focused on financial and economic issues. Questions such as the transparency of financial transactions and liberalization of cross-border flows were dominant themes. By easing the constraints of national boundaries, the hope was that a variety of interests would benefit and different countries would gain from free trade.[35]

As production shifted across national boundaries, economic efficiency would be improved. If a particular country had national advantages due to location, wage structure, or natural resources, it would find a niche for particular products or services that would be valuable for the international community. Jobs would be created locally, and people around the world would be able to import those products or services at a cheaper cost. Countries would segment into importers or exporters, or service providers or manufacturers, and the entire arrangement would maximize efficiency in the global marketplace.[36]

To some observers, economic globalization is an unmitigated good. Economists sold the idea to policymakers on the basis of market efficiency and economy of scale. Developing nations were promised markets for their products. Industrialized nations looked forward to cheaper consumer products that also would provide jobs for low-skilled workers in poor countries. In short, globalization would be helpful for global trade and a growth engine for the entire world.

But not everyone is enamored with the supposed virtues of globalization.[37] Critics predicted a "race to the bottom" that would harm union workers in industrialized nations, create loopholes in environmental and safety regulation, and perpetuate poverty in developing countries. Rather than boosting efficiency and productivity, globalization would fuel inequality and injustice around the world.[38] Political scientist Philip Cerny complained that globalization would create a democratic deficit. The power of international governance bodies threatens the autonomy of individual nations and undermines democratic control by ordinary citizens.[39]

Although evidence does not support all the predictions of this race to the bottom, there is little doubt that inequality has increased and the gap between north and south has widened. Globalization has allowed businesses to play countries off against one another. Places that offer low wages or weak environmental and safety enforcement are seeing job growth, generally at the expense of high-wage areas. Countries such as China and India are experiencing tremendous growth rates due to their inexpensive sources of labor.[40]

In debates over globalization, though, much of the focus has remained on its economic or political ramifications. Old globalization evokes many reactions, but conflict centers on matters such as job creation, wage growth, trade patterns, and income inequality. Most of the arguments rest on economic reasoning about how global trade is affecting the material fortunes of particular nations, industries, or segments of the labor force.

These consequences are important, but a new globalization is emerging that involves moral reasoning, and emotional debates over the effects of boundary lowering on the sanctity of life, privacy, and security. Controversies over stem cell research, cloning, chimeras, genetically modified food, and in vitro fertilization, for example, focus on whether technical advances in these areas have gone too far and are endangering fundamental human values.[41] Similarly, the emergence of supercomputers, the Internet, and a networked society raises doubts about privacy and confidentiality in a wired world. In the post-September 11, 2001, world of international terrorism, there is renewed concern regarding whether reducing border barriers creates security problems that endanger national security.[42]

Each of these debates is fundamentally different from arguments about old-fashioned economic liberalization. Rather than raising questions regarding efficiency, economy, income generation, and market segmentation, the new globalization must deal with issues involving moral and ethical undertones that are quite difficult to compromise. Biotech globalization involves a wide range of actors outside of government. The rise of fundamentalist religious forces in many parts of the world adds a dimension to conflict that is emotional and intense. The state in secular countries is not powerless in these arguments, but civil, religious, and nongovernmental organizations have significant ability to block action and fight for ethical perspectives on genetic engineering.

All this makes the future of globalization much more conflicted, emotional, and tenuous than would appear to be the case under the old globalization. Instead of negotiating free trade rules, countries now worry that cross-border trade and traffic will spread illnesses such as Severe Acute Respiratory Syndrome (SARS) or Acquired Immunodeficiency Syndrome (AIDS), subject computer users to viruses and spam, or import values that are abhorrent to particular segments of the world. As political scientist Samuel Huntington aptly put it, a "clash of civilizations" in the new globalization pits competing values and world orientations at center-court in discussions of international commerce.[43] The intensity associated with different worldviews complicates the development of biotechnology and helps to explain why various countries have adopted such different approaches to handling these kinds of issues.

Chapter 2

Science-Industry Collaboration

In 1982, the Animal Breeding Research Organization of Scotland had a serious fiscal problem. The British agency in charge of funding its research faced a budget shortfall that led to spending reductions for recipient institutions. Rather than lay off scientists from its basic research team, the Scottish scientific organization decided to generate new revenues by commercializing its products. Renamed the Roslin Institute, it created a privately funded company, PPL Therapeutics, in 1987 to bring products to the marketplace. Within a decade, PPL scientists were part of the research group that produced Dolly, the world's first sheep to be cloned from adult cells. The company went on to pioneer new treatments for several diseases and partner with leading pharmaceutical companies.[1]

This close collaboration between scientists and industry is not unusual in many countries. France, one of Europe's most state-focused nations, offers its scientists at public research institutes up to six years of sabbatical leave to develop private companies. During this gestation period, these individuals retain their paid privileges as civil servants. They only have to choose between their public and private sector jobs at the end of the six years.[2]

American professors also have become prolific in creating private companies that commercialize their research. The full extent of these science-industrial collaborations was documented when an article about scientific papers published in major biomedical journals discovered that one-third of the authors held a financial interest in the research being published. This ranged from equity and patent holdings to serving on boards of companies linked to the research.[3] These kinds of financial arrangements have become a major source of revenue for universities.

There are material ties between science and industry in other lands. After the South Korean scientist Woo Suk Hwang claimed to have extracted stem cells from human embryos (research that was found later to have been fabricated), Korean Air gave him financial support for his research and free air flights for a period of ten years.[4] In China, the line between state enterprises and private companies has become blurred as a number of public agencies have launched for-profit, biotech firms.[5] In 2000, Singapore created a private firm, Lynk Biotechnologies, comprised mostly of researchers from the National University of Singapore. Currently, the country has more than 30 "bioventures."[6]

In this chapter, I look at the science-industrial complex in biotechnology. There are a number of trends that have brought science and industry much closer together: the reshaping of the public sector in many countries that has diminished the role of government; the use of corporate partnerships and start-up companies in higher education; the power of technical expertise; the emergence of "country-shopping," "scientist-buying," and medical tourism as ways to leap forward; and the prevalence of contraband technology in global trade. I examine these trends to see how the science-industrial complex emerged and what this has meant for biotech policy. I argue that in fundamental respects, the balance between public, private, and nonprofit organizations has shifted dramatically in favor of non-state actors.

Reshaping the Public Sector

The prevailing public philosophy in many countries over the past few decades has been deregulation, tax cutting, privatization, and transferring public sector responsibilities to nongovernmental organizations.[7] Ever since the emergence of Thatcherism in Great Britain in 1979 and Reaganism in the United States in 1980, a number of governments have altered their approach to public policy. Rather than looking to the public sector to finance major activities, politicians have recast the scope of government and chosen to rely extensively on industry, universities, trade associations, and nonprofit groups for various services.[8]

By doing this, reformers hoped that societies would become more efficient and entrepreneurial. Antigovernment rhetoric decried the inefficiency, ineffectiveness, and low productivity of the public sector, while simultaneously praising nongovernmental organizations for their skill at creativity and innovation. Politicians campaigned against bloated federal bureaucracies and ran on platforms proclaiming the need to reduce taxes, sell off state assets, and trust the power of market competition to spur human productivity.

In America, President Ronald Reagan (and later the two Presidents Bush) preached the virtue of downsizing the governmental sector. In 1981, Reagan pushed a major tax-cut program through Congress that reduced personal income taxes by 25 percent.[9] This across-the-board tax reduction was billed as a way to jump-start an economy in recession following the oil shocks of the 1970s and the emergence of "stagflation," which is to say, the combination of high unemployment with high inflation.

By providing more after-tax dollars for businesses and ordinary citizens, Reagan (and the Bushes) argued that people would have extra money for discretionary spending and that the ability to retain more of one's earnings would spur productivity. Dubbed "supply-side" economics, this reasoning became the new mantra of conservatives around the world. When the economy recovered following the tax cuts, Reagan coasted to a landslide reelection that included the support of some Democrats.[10]

Congress also deregulated public control of a number of key sectors of the economy, and privatized some government functions.[11] Democratic president Jimmy Carter deregulated the airlines and natural gas in 1978 and trucking and financial institutions in 1980. Reagan followed by deregulating radio and television in the 1980s, while President Bill Clinton successfully pushed for telecommunications deregulation in 1996.

When high budget deficits emerged during Reagan's administration, the chief executive pressed to reduce government spending. Rather than admitting that tax cuts did not boost revenue as much as had been projected, the president continued his antigovernment rhetoric and insisted public sector spending had to be cut.[12] Since he wanted to strengthen the military and spend more on defense, much of the resulting fiscal pain fell in nondefense areas, such as social services, infrastructure repair, and research and development.

Consistent with the rhetoric of smaller government, Reagan encouraged the private sector to assume responsibility for providing basic services. Banks, insurance companies, and savings and loan associations had more leeway in offering financial products. Welfare cuts meant that nonprofit organizations had to provide help for those not covered by government. And in the area of research and development, industry and universities were prodded to provide additional support for scientific innovation.

Around the same time that Reagan and his allies were pushing for tax cuts and deregulation, the British prime minister Margaret Thatcher stressed the need for a new entrepreneurial spirit in Great Britain.[13] The nation should not rest comfortably on generous social insurance and government resources, she said, but rather its businesses and scientists ought to think differently. Markets should be opened up, she said, and state assets such as British Airways and British Steel should be privatized and made to operate more efficiently.

Although this move was controversial in a society that long had provided generous social insurance and seen the government playing a constructive role in society, Thatcher's antigovernment rhetoric fundamentally altered civic discourse and public policy. Taxes were cut, public companies were sold to the private sector, and government control over significant parts of business and the economy was deregulated.[14]

It took a while for the Thatcher Revolution to filter down to higher education, but British institutions of higher learning were not immune to this change in public philosophy. The financial squeeze on public spending that resulted from tax cuts led British universities to seek new sources of revenue. Similar to their American counterparts, scientists in Great Britain sought closer ties to industry, and the commercialization of scientific discoveries became more central to the academic mission.

The fall of communism and the rise of globalization in the 1990s sealed the triumph of capitalism and the ascendancy of "the market" as the dominant governing idea. The breakup of the Soviet Union and the delegitimization of socialist ideology removed one end of the pro-statist political spectrum and transformed debates about the public sector. Rather than argue over whether the public or private sector was the best way to serve people, market forces were trumpeted as the most efficient. This shifted public discussion toward how to use market competition to serve human needs, reform schools, and administer health care, as opposed to whether the market was a viable means to allocate resources and decide government priorities.

Even liberal politicians embraced the rhetoric of investment and the marketplace. Democratic president Bill Clinton became a leading advocate of private-public partnerships that sought to marry the efficiency of the private sphere with the ambitions of the public sector. For example, welfare reform ended the notion of a long-term government entitlement program for the poor. Instead of being able to remain on public assistance indefinitely, government aid was limited to five years, with the expectation that employment training programs and education would move recipients from the public welfare rolls to the private labor force.

In the international sphere, globalization was promoted as the way to link nations together more closely and more efficiently. Economic barriers to trade between nations were reduced, and global commerce was emphasized. Nations that had natural advantages in some areas specialized in those products, and service delivery and manufacturing migrated to areas of greatest efficiency. Market forces decided which countries dominated particular areas. This new global competition was designed to ease trade and allow all countries to deliver benefits to their citizens.

The impact on biotechnology of the public sector's diminished role was immediate and direct. Organizationally, science moved closer to industry in the emphasis on commercialization. The previous period in higher education when basic knowledge was valued as an end in itself gave way to an era of applied and practical research. Academic knowledge was directed to commercial products, and the resulting royalties and patent income enhanced the universities' ability to support scientific innovation.

In addition, universities came to be seen as reservoirs of expertise for political decision making. Increasingly, the public sector turned to technical experts, advisory councils, and professional associations to make government policy. Priorities and ethical dilemmas surrounding new initiatives were debated by experts outside the public sector. This elevated technical expertise to a new high in civic discourse.

The impact of all of this on funding was quite discernible. Government would provide support for basic research and infrastructure, but nongovernmental organizations were expected to apply knowledge and be the major scientific innovators. Since the public sector was not viewed as very creative or efficient, this outcome was consistent with the pro-market rhetoric of leading elected officials and the prevailing ideologies of the time.

By the turn of the twenty-first century, the triumph of market ideals had produced a sharp decline in government support of research and development in many countries around the world. Across the 30 countries of the Organisation for Economic Co-Operation and Development, the percentage of research and development provided by government dropped from 44.4 percent in 1981 to 28.5 percent in 2000. By the beginning of the new century, government support for research and development had fallen to 20 percent in Japan, 25 percent in Korea, 27 percent in the United States, 29 percent in Great Britain, 31 percent in Germany and Canada, and 39 percent in France. Meanwhile, industry support comprised 72 percent of research and development in Korea and Japan, 68 percent in the United States, 66 percent in Germany, 52 percent in France, 49 percent in Great Britain, and 42 percent in Canada.[15] The era in which government provided the bulk of financial support for scientific research was over.

Corporate Partnerships and Start-Up Companies in Higher Education

In this new era of market capitalism and scarce public resources, the lesson for universities was clear. Higher education had to become entrepreneurial and seek money from commercial revenue sources. No longer could colleges and universities rest on their past successes in providing educational opportunities and thereby furthering the American Dream. Rather, they needed to raise money, collaborate with private companies, and develop for-profit wings of their own enterprises.

In many respects, universities succeeded beyond their wildest hopes. The number of university-owned patents has risen from 95 in 1965 to 3,200 in 2000 and over 8,000 in 2004.[16] Revenues from outside sources are up as well. According to industry surveys, American universities earned $1 billion in patents and licensing fees in 2002.[17] Over the years, several schools have hit the jackpot on individual discoveries. For example, a technique for transferring genes from an African clawed toad to bacteria (the basis for genetic engineering) earned $300 million apiece for Stanford University and the University of San Francisco, as well as for inventors Stanley Cohen and Herbert Boyer. The Gatorade drink has generated $94 million for the University of Florida. Florida State University has received over $200 million in fees based on the work of scientists who helped produce Taxol, an anticancer medicine.[18]

Academic research support provided by commercial firms in the United States has risen from $236 million in 1980 to $2.234 billion in 2001.[19] The commercial sector now provides just under 10 percent of the money for university research. At some schools, however, this figure is considerably higher. For example, private industry provides 48 percent of the research budget at Alfred University, 32 percent at Tulsa, 31 percent at Duke, 22 percent at Lehigh, 21 percent at the Georgia Institute of Technology, 20 percent at MIT, 16 percent at Ohio State, and 15 percent at both Penn State and Carnegie Mellon.[20]

Universities have pursued corporate partnerships with great enthusiasm. In exchange for financial support of academic units, businesses have been given opportunities to commercialize academic knowledge. One of the most infamous examples of this collaboration between science and industry took place at the University of California, Berkeley, with its department of plant and microbial biology. A private company called Novartis paid $25 million to gain first rights to commercialization. Not only would the firm have the opportunity to license university discoveries, it would occupy 40 percent of the seats on a department committee that allocated research funding.

This alliance generated tremendous controversy over possible conflicts of interest.[21] Critics decried the intrusion of commercial values into the academic enterprise. An external review suggested the university had gone too far in sacrificing academic freedom. After an opponent of the initiative was denied tenure by the university, a judge overturned the tenure denial and awarded money to the plaintiff. In the end, though, these corporate partnerships became popular at many universities in the United States.

Start-up companies developed by professors have become key ways for universities to develop new revenues. With changes in U.S. patent law in 1980 through the Bayh-Dole legislation, universities discovered they could make large amounts of money by patenting scientific discoveries. New revenues would be tapped by encouraging scientists to develop for-profit enterprises in which the university owned an equity stake. By 2001, 886 new private companies had been started by American professors.[22] For example, the Stanford scientist Irving Weissman created a company called StemCells Inc. to inject human stem cells into mice and see whether these cells aided in fighting human diseases.

As a further sign of the corporatization of higher education, many U.S. universities built business parks and research incubators designed to nurture commercial enterprises.[23] Among the 174 universities that constructed such facilities, the most famous is the Research Triangle Park linking Duke University, the University of North Carolina at Chapel Hill, and North Carolina State University. This area has been a boon to research and development in a variety of areas. It has fueled economic growth, community development, and scientific innovation.[24] Similar facilities have been developed by Stanford MIT, Texas, Stanford, Chicago, and Harvard, among others.[25]

Research on these university incubators demonstrates that higher education has become a major regional booster for technological innovation. Not only are these facilities associated with higher economic growth, but there is also evidence that these parks produce spillover effects on new start-ups and that they generate additional research and development spending. Cities and states are now quite eager to develop these research facilities.[26]

Other nations have developed close ties between science and industry. In Great Britain, for example, Microsoft has sponsored labs at Cambridge University for the promotion of new technology. Cambridge, indeed, is home to six "knowledge parks" designed to jump-start innovation. Germany has witnessed alliances between higher education and leading companies such as Hoechst, and has seen Munich become home to many new biotech collaborations. Sweden has spun off more than 90 companies from Chalmers University alone.[27] Japan, India, China, and Korea have also seen a burst of science and industry collaboration. Science parks for commercial collaboration have spread around the world. Public-private partnerships are thought to be an ideal way to benefit both the private and nonprofit sectors. Through them, academics gain access to needed financial resources, while the private sector benefits from knowledge specialists who spur new discoveries.

The Power of Technical Experts

The importance of innovation and the rise of science-industry collaborations have elevated expertise to a special status in many societies. Technical experts are employed in many countries to decide scientific controversies, serve as referees for peer review, and help shape future priorities for funding agencies. Indeed, with the proliferation of specialized knowledge among nongovernmental organizations, many agencies rely extensively upon outside experts to formulate national policy and allocate scarce financial resources.

Even though many nations have public sectors that are investing significant amounts of money in biotechnology, this does not mean that government dominates decision making in this field. As the biotech analyst Sheila Jasanoff has pointed out, technical experts have become the "fifth branch" of government in many places. So-called knowledge specialists have carved out major policymaking roles for themselves in various countries through advisory councils, nonprofit groups, and professional associations.[28]

Many nations use advisory groups to shape judgments about scientific priorities. In the United States, for example, expert panels advise the National Science Foundation and the National Institutes of Health. Since 1995, the American president has had an advisory committee on bioethics that offers guidance on cloning, stem cell research, and other new initiatives. In Great Britain, the government establishes blue-ribbon committees composed of recognized experts to issue papers on major controversies. Such a committee, headed by Oxford University philosopher Mary Warnock, produced a report known as the Warnock Report in 1984 that recommended specific guidelines for the conduct of embryonic research.[29] Another body, the Nuffield Council on Bioethics, has issued reports on food and biotech guidelines. The German chancellor has a National Ethics Council drawn from different fields that advises her on major initiatives.

Most of these experts work either in universities, nonprofit organizations, or other kinds of nongovernmental programs. There has been a tremendous growth in nongovernmental organizations throughout the world. These bodies have great power and shape much of what happens in biotechnology. Their edicts on ethics and resource priorities receive considerable press attention and are debated at the highest levels of government.

While presented to the public as a nonpolitical and technocratic approach to policymaking, the reality is that experts bring particular values to decision making and represent various types of social and economic interests. Some agencies have been accused of being "captured" by leading scientists. Rather than being run by administrators, they are seen as having lost effective control to outside experts.[30]

This power is a legitimate concern in policy areas having low visibility and low conflict. It is in these areas that experts exert their strongest power. Areas such as Internet regulation and agency funding allocations typically are "inside" political games where scientists use their specialized knowledge to shape public policy. Citizens and elected officials are not aware of or concerned about these choices. Instead, key decisions are delegated to knowledge specialists.

In areas having higher visibility and conflict, though, expert power is constrained by broader political forces. For example, in areas that evoke great controversy and attract considerable media attention, such as cloning and stem cell research, experts generally have less power, and elected officials, media reporters, and the general public (through polls or election results) exert more control.

The high visibility of these policy disputes and the prevalence of political partisanship, ideology, and religious values reduce the ability of experts to bring their specialized knowledge to political debates. Rather than being decided out of the public spotlight based on technocratic criteria, these matters are subject to party conflict, ideological disputes, and ethical values. This broadens the scope of conflict in the biotech area and involves a wider range of political participants.

As the following chapters will show, some biotech decisions, such as in vitro fertilization, assisted reproductive technology, and genetically modified foods (except in Europe), attract little public attention. Experts make major decisions, and the science-industrial complex is very influential. In other areas, such as cloning and stem cell research, controversies over the ethics of biotechnology elevate public awareness, generate press coverage, and draw in elected officials. This reduces expert influence and weakens the ability of the science-industrial complex to dominate decision making.

The Emergence of "Country-Shopping," "Scientist-Buying," and Medical Tourism

The mobile nature of knowledge in the contemporary period has freed technology innovation from some of its prior geographical constraints. Unlike heavy industry or specialized manufacturing, which are somewhat bound to particular areas by the proximity to raw materials or the location of existing plants, biotech research can move around rather freely. Subject to rules imposed by particular locales, research involving stem cells or genetically modified organisms can take place in a variety of places. Indeed, with easy flights from country to country and electronic communication via e-mail and the Internet, scientific knowledge is more movable than ever before.

This mobility results in some scientists, doctors, and patients "shopping" for countries with the most favorable rules and resources. Rather than accepting government restrictions on what they can study, scientists and funding organizations migrate to places with few government rules or lax enforcement of formal regulations. Small countries in out-of-the-way places are often eager for the investment and economic development that can accompany scientific research and medical treatment. Even though the fiscal payoff of these investments is sometimes disappointing, some countries continue to push for these kinds of development projects.

As an illustration, the Caribbean island nation of St. Kitts represents a place that has established itself as friendly to researchers. One of the hot topics for the biotech industry has been implanting stem cells from one species into another as a way to model health consequences. Called "chimeras" (or new cross-species organisms named after the mythic Greek creature), these experiments are controversial in some countries.

St. Kitts has a burgeoning industry based on chimera experimentation. It is attractive for chimera research because monkeys are in abundant supply there and the government imposes few rules on scientists. There have been no signs of animal rights activists who have terrorized researchers in the United States and Europe by breaking into biotech labs, freeing laboratory animals, and destroying scientific equipment. Reporters rarely travel to this area to inspect research facilities.[31]

Taking advantage of the geographic isolation, Professor Eugene Redmond of Yale University established the St. Kitts Biomedical Foundation in 1982 to support research that injects human brain cells into monkeys. His goal was to see whether the new brain cells would multiply, produce dopamine, and therefore help patients with Parkinson's disease recover from their illness. Armed with the freedom to undertake this controversial research, Redmond's team has identified ways in which medical personnel can better target pathways for treatment.[32]

Due to ethical concerns over chimeras, Canada and Japan have placed major restrictions on this type of research. There is obvious fear over the propriety of cross-species transplantation, and an ethical dilemma over whether scientists should be able to imbue monkeys or other species with human characteristics. In contrast, other countries, such as the United States, India, and China, are not very restrictive in their approach to this research. There are fewer philosophical objections to this kind of work and hope that chimeras will lead to new treatments for serious illnesses.

"Country-shopping" gives scientists great power over nation-states. Rather than having to accept research restrictions or deal with protests from animal rights activists or media reporters in developed countries, scientists migrate to locales that are close to natural resources and far from political scrutiny. Other than regulations imposed by their home universities, they don't have to worry about restrictive research protocols or human subjects committees. They can pursue new research ideas with great autonomy.

Scientific funding follows research, and in some cases it too has migrated to geographic locales with few research restrictions. For example, after many Western nations discouraged research on stem cells, the American-based Juvenile Diabetes Research Foundation began funding this kind of research in Singapore. Sponsoring a Spanish scientist, Bernat Soria, who faced research restrictions in his home country, the organization provided a generous grant that allowed him to continue work that was impossible in both Spain and the United States. When asked about this arrangement, the group's scientific officer justified the move by describing Singapore as a place where "there is excellent science, a good environment, and really strong support for work that can't be done in the U.S."[33]

"Scientist-buying" is another strategy for economic development. Raiding another country's expertise is a way for nations to leap ahead scientifically. Singapore, for example, has adopted a general policy of bringing well-known experts from other lands to their nation. In 2000, it decided to spend $2 billion on what it called its National Biomedical Science Strategy. This initiative provided funds to recruit scientists from other countries and create tax incentives for commercial firms. Researchers can get up to three years of leave time if they start a new business. Scientist-buying is a mechanism by which Singapore hopes to leap forward quickly in the biomedical area.[34]

The commitment helps the country of 3 million residents attract nearly half a billion dollars of foreign biomedical investment each year. Several hundred scientists from the United States, Great Britain, Japan, and Australia have accepted Singapore's offer of new employment. Among them was biologist Alan Colman of PPL Therapeutics, the firm that created the cloned sheep Dolly. He joined ES Cell International to commercialize academic work completed at Singapore's National University Hospital.[35] An American scientist with a lab in Singapore, Ian McNiece, described the research advantages of Singapore over the United States as freedom from "the whims of American politics" and "lawsuits intended to block research."[36]

In an era of global travel, some businesses have combined tourism and experimental therapy into a new industry known as medical tourism. This practice allows patients seeking new health care treatments to fly to a country, get medical assistance, and tour exotic locales. An Argentinean company by the name of Plenitas, for example, advertises medical tourism through its website, www.Plenitas.com. This site discusses how individuals in need of first-class medical care can travel to Buenos Aires to receive plastic surgery, hair treatment, fertility therapy, ophthalmology, and other cutting-edge forms of surgery and health care. The company promotes experimental techniques for regenerating arteries through skin cells as well as loans to pay for medical services. The multilingual site advertises its services in Spanish, English, and German.

These are just a few of the cross-border enterprises that have sprung up in recent years. The mobility of knowledge, labor, and money in an era of globalization gives scientists freedom that would have been impossible in earlier time periods. Just as companies shop around for the cheapest manufacturing labor or lowest taxes and patients search for experimental medical treatments, scientists look for countries with weak rules or lax enforcement.

With the number of international coauthorships on American scientific papers at an all-time high of 25 percent (up from 10 percent in the late 1980s), scientists and their research funding are being liberated from geography. They are global players who travel frequently and maintain international knowledge networks.[37] They attend international conferences and share knowledge with peers from other lands.

This geographic flexibility adds new power to the science-industrial complex. Rather than being subject to government restrictions, the knowledge elite has been freed to pursue scientific innovation that is opposed in their own homelands. The ability to move around has given biotech experts a degree of autonomy that early capitalists would have envied.

Contraband Technology

The free flow of technology across national borders is a hallmark of globalization. Indeed, increases in global trade represent one of the goals of new financial regimes that lower transaction costs and increase transparency across different countries. The hope is that nations will increase international commerce, lower trade barriers, and thereby create more efficiency in the overall system.

Even in a time of global trade, however, there remain constraints on various transfers and transactions. Countries often restrict research and trade in areas it views as sensitive, ethically suspect, or contrary to national security requirements. For example, trade in supercomputers that have military applications is banned with countries that are considered terrorist threats. Some countries such as China use "filters" to keep certain text messages and Internet sites away from domestic consumption. In the biotech area, genetically modified foods may be acceptable in one country but not in another, and therefore limited from global exchange. Some nations may limit certain types of commerce for religious reasons. Still other nations may want to keep particular products out in order to protect home-grown industries.

Despite such bans and restrictions, however, it is nearly impossible to prevent people from trading in contraband technology. Greed, corruption, or lax enforcement allows products and services to move around the world, even if there are concrete restrictions against trading those kinds of commodities. In some cases, government officials may pay lip service to these bans, while still trading prohibited materials with willing buyers. In other cases, bribery allows materials to leak through borders. Or there may be lax enforcement of national or international rules that facilitates the transfer of goods. One way or another, contraband products move around the world.

A recent example of this is the discovery of genetically modified rice in China. China bans the use or importation of altered rice. Because of the centrality of rice to its 1.3 billion citizens and its interest in protecting homeland producers, the Chinese government has been very careful to regulate rice experiments and will not approve the use of pest-resistant rice until it is absolutely certain the food is safe for consumers.

It was surprising, then, when genetically modified rice appeared in markets around the Hubei province in direct violation of government laws banning its sale. The environmental group Greenpeace, which opposes the use of genetically modified foods, tested the rice and confirmed it had been genetically altered. Even though government officials had not approved the sale, an investigation revealed that the modified rice was being sold at a store run by Huazhong Agriculture University in Wuhan. Government-run markets in the area were also found to be selling genetically modified rice.[38]

These findings show the power of contraband technology and the difficulty of enforcing laws outlawing trade in illegal commodities. If a product has a market and people are interested in buying or selling it, government officials will find it difficult to control trade. History has shown that humans are endlessly creative when it comes to circumventing formal prohibitions.

A similar problem also arose in Japan, this time involving the importation of genetically modified corn. Even though the Japanese government has strict policies against importing genetically modified grains, traces of Bt-10-contaminated corn were found at a Japanese port in 2005. Three different cargo ships had entered the port carrying the gene-altered food. The discovery surprised government officials, who threatened to penalize Syngenta, the Swiss biotech company that had produced the corn.[39]

Because governments have difficulty limiting international commerce, biotech is an area that is very difficult to regulate. Even when governments have formal prohibitions on the books, it does not mean science and industry cannot find ways around the restrictions. Indeed, it is the mobility and decentralized nature of biotech that present officials with some of their greatest challenges. With biotech's great reliance on experts from outside the government and difficulty controlling the knowledge industry, it has become nearly impossible for the state to exercise significant oversight. As discussed later, this has tremendous ramifications for the ethics and dynamics of decision making in countries around the world.

Chapter 3

In Vitro Fertilization and Assisted Reproductive Technology

In vitro fertilization represents a biotech procedure that largely has been privatized between patients and medical personnel. Even though it bears directly on procreation and is fundamental to the creation of life, it is not controversial in most places around the world. It does not evoke protests that doctors and patients are "playing God" or creating Frankenstein monsters. Activists do not sit outside infertility clinics demanding patient boycotts and an end to the therapy. Countries either have passed no legislation to regulate practices in this area, or they have minimal rules that allow scientists, private companies, and patients to make major decisions on their own, unimpeded by governmental restrictions.

In this respect, assisted reproduction technology exemplifies the most permissive example of the science-industrial complex. Rather than restricting private decision making, most nations allow patients and medical personnel the freedom to make their own decisions on fertility treatment. Governments impose few restrictions on the choices of married couples and trust medical personnel to define ethical practices. The most serious limits on access are personal and financial, rather than political or legal.

The major exceptions to this rule are the predominantly Catholic countries of Italy and Ireland, and the special case of Germany following World War II. Both Italy and Ireland have comparatively tough restrictions on fertility treatments and assisted reproductive technology. The combination of a large Catholic population, an activist and politically powerful church, and traditional social values has produced rules (especially in Italy) that are more severe than in most other countries around the world. In the case of Germany, the government enacted strict ethics rules for assisted reproduction in the aftermath of World War II and its notorious Nazi medical experiments. This legacy continues to color its biotechnology policies to the current day. The study of policymaking in these nations, therefore, represents an interesting opportunity to examine contrary cases, and how religion and culture allow the state to regulate in vitro fertilization in some locales.

Overview of In Vitro Fertilization and Assisted Reproductive Technology

In vitro is Latin for "in glass," meaning life that is generated outside the body through artificial means.[1] It refers to assisted reproductive technologies developed by scientists and doctors designed to address fertility problems defined by the Center for Disease Control as "not being able to get pregnant after trying for one year."[2] In the United States, there are around 60 million women of reproductive age, generally considered to be those from 15 to 44 years old. Of these women, 2 percent report they have sought fertility assistance in the last year, and 15 percent did so at some point during their lives.[3] In addition, around 30 percent of American adult men are considered infertile.[4]

Infertility varies by specific country. In Great Britain, for example, it is estimated that 10 percent of women are infertile, while in some African nations, the number runs as high as 30 percent. The larger percentage of infertility in African countries is due to poverty, poor levels of medical treatment, and a high incidence of sexually transmitted diseases.[5]

In postindustrial societies, fertility problems arise when couples delay childbearing until later years. Infertility rises with age so the decision to defer children increases the difficulty of getting pregnant. Studies estimate that the odds of being infertile increase from 8 percent of all women between 19 and 26 years old, to 13–14 percent of those between 27 and 34 years, and to 18 percent for those between 35 and 39 years old.[6]

Treatments vary by individual need, but the most common approaches include injecting sperm into an egg, transferring a donor egg, embryo freezing, assisted hatching, sperm storage, artificial insemination, or drug therapies to stimulate egg production. Generally, in vitro techniques involve taking eggs and/or sperm cells out of the body, combining them through artificial means, and reinserting the impregnated egg in the woman. According to government figures, around 25 percent of these interventions produce babies.[7]

The vast majority (85 percent) of women with fertility issues do not seek treatment. In vitro fertilization is expensive and time consuming, and most insurance companies do not cover the procedure (although 15 American state governments have passed legislation requiring insurers to cover fertility treatments for women under a certain age, anywhere from 40 to 46 years old).[8] With costs in Western countries ranging from $5,000 to $12,000 per procedure and many women requiring four treatments (or a total of up to $20,000–$48,000 per baby), many of those affected do not utilize this elective therapy.[9] If the treatment is not beyond their financial means, couples are often discouraged by the physical and emotional stresses that accompany assisted reproductive therapies.

However, the rest undergo some form of fertility assistance. Scientists and commercial clinics have collaborated to produce a range of new medical therapies. These new technologies are easily transferred from clinic to clinic and across national boundaries. The result has been a globalization of assisted reproduction. Doctors exchange information on success rates and compete for patients, sometimes even across state or national borders.

Infertile couples living in nations with strict limitations on the use of procreation technologies can opt for "medical tourism" and travel to another state or country for treatment. Except for the most authoritarian regimes that limit freedom of movement, there are no rules preventing patients who desire advanced medical techniques from receiving treatment in places where there are fewer restrictions. For those who can afford to pay for assisted reproduction, this option of mobility undermines governmental attempts at regulation. The practices of medical tourism and "country-shopping" illuminate the difficulties for those worried about the ethics of reproductive technologies to enforce their values on other people.

The first and most famous example of a test-tube baby was British-born Louise Brown. It was in the 1970s that John and Lesley

Brown approached medical doctors in Great Britain with a problem. They had been unable to conceive for a period of nine years due to blocked fallopian tubes. After being referred to Doctors Patrick Steptoe and Robert Edwards, the couple consented to a new experimental procedure. Physicians removed an egg from Lesley Brown, combined it with her husband's sperm, and surgically implanted the fertilized egg in the woman's uterus. On July 25, 1978, Louise Brown was delivered as a healthy baby via cesarean section.[10]

From this pioneering effort arose a multi-billion-dollar industry that made conception possible for many couples around the world. In the United States, there are 400 in vitro fertilization clinics that produce 75,000 births annually. Europe has 740 fertility centers. Even the heavily Catholic country of Ireland has nine medical facilities to assist those with infertility problems. Around the world, it is estimated that two million in vitro babies have been born since 1978.[11]

However, the risks of in vitro techniques remain high. As part of the procedure, doctors often transfer two or three embryos into a woman's uterus. This increases the odds of twins or triples, respectively, and elevates the probability of miscarriages, birth defects, and other medical complications. In vitro health risks rise significantly for women in the late 30s or 40s at the time of their procedures.[12]

The American Experience

In the United States, there is little government regulation of in vitro fertilization or assisted reproductive technology. Medical clinics are not certified for this purpose and there are no quality control inspections. Patients do not have to sign informed-consent forms nor is there any required counseling for patients prior to treatment. Of the rules that are in place, many of them are at the state level, and there is wide variation from place-to-place in the guidelines that exist. Louisiana has some of the toughest restrictions due to its legal definition stipulating that embryos have the same legal rights as humans, while California, New York, and Illinois impose few constraints on assisted reproduction.[13]

In 1992, Congress passed legislation entitled the Fertility Clinic Success Rate and Certification Act, designed to address concerns over clinics' lack of standardization of treatment success rates. Patients had difficulty comparing pregnancy rates across a range of different treatments and facilities because physicians reported data in conflicting ways. This federal legislation mandated a common protocol for reporting the effectiveness of various therapies, with the goal of helping patients understand clinic performance, and therefore, create more accountability in the fertility industry.[14]

But the law did not regulate the manner in which in vitro techniques were performed. Indeed, the legislation explicitly prohibited interference with or oversight of clinical practices. Policymakers believed that doctors are the experts in assisted reproductive technology, and the public sector has no reason to regulate these kinds of medical practices.

As pointed out by ethicist Cynthia Cohen, the absence of strict in vitro regulations is unusual because most medical facilities and other procedures involving surgery face strict rules. Clinical labs are regulated, as are hospitals, outpatient clinics, and blood banks. Hospitals, for example, are accredited through state organizations. There also are professional associations and private bodies that oversee physician conduct.[15]

The absence of medical regulations on in vitro fertilization demonstrates the factors that create autonomy for particular biotech areas. One reason for the lack of government oversight is because medical professional associations have lobbied to prevent regulation of their practices. By working hard to exempt their domain from medical oversight, in vitro specialists have created an unusual level of independence for their field.

Another reason for weak governmental oversight of these procedures is the strong public opinion in support of assisted reproductive technology. Unlike other biotech procedures where public attitudes are polarized by ideology, religious beliefs, or political party, opinions regarding in vitro fertilization are favorable to the technology and the broader mission of childbearing by married couples.

A 1978 Gallup national public opinion survey in the United States, conducted directly after Louise Brown's test-tube birth, found that 60 percent of Americans favored in vitro fertilization, 28 percent opposed it, and 13 percent had no opinion. In addition, 53 percent indicated that they would be willing to try in vitro methods if they wanted a child and were unable to conceive, while 35 percent would not.[16]

Since that time, Gallup has not polled the in vitro question, which is a sign of its relative noncontentiousness in the United States. The Gallup organization has polled other issues surrounding assisted reproduction, such as use of surrogate mothers. One survey conducted in 1987, inquired whether individuals would consider having a child by a surrogate mother if the individual wanted a baby but could not conceive because of infertility or other health reasons. In this poll, 35 percent indicated they would be willing to rely on a surrogate mother to produce a baby.[17]

A few years ago, pollster John Zogby undertook a general survey of attitudes on in vitro among American Catholics. One might expect Catholics to be more cautious about in vitro given Church teaching against its practice. However, even among this group, there was substantial support. When asked whether they agreed or disagreed with the church's stance that in vitro fertilization was wrong, 50 percent said they disagreed with this church teaching, while 44 percent agreed with it.[18]

In recent years, there have been some efforts at the state level to establish legal rules on aspects of assisted reproduction. It is primarily the ethical complications surrounding private market contracts for the donation of eggs that has led to state laws regulating such actions. For example, about half of American states require mandatory testing of donated sperm for human immunodeficiency virus (HIV). Around half of the states also require that a husband give his written permission before a married female can accept donated sperm. Some states ban the sale of human embryos. Others have placed legal restrictions on intermediary "baby brokers" who match eggs with sperm. Still others have sought to define parental rights when donor sperm is involved.[19]

In 2003, there was some trepidation within the fertility industry when the President's Council on Bioethics began to discuss possible guidelines. The fear was that the Bush administration and their allies in the "religious right" might curtail the laissez-faire tradition that has existed for assisted reproduction in the United States. Industry officials worried that new rules grounded in a conservative moral philosophy would move America toward stricter reporting requirements and outright regulation, rather than continuing the hands-off approach.

Such fears were unfounded, however, as lobbying by fertility experts led the council in its 2004 report to stress the need for "enhanced professional guidelines," rather than mandatory rules or specific forms of oversight.[20] The result is that, despite having a national government controlled by conservative Republicans who campaigned on moral issues and are supported by the religious right, the United States continues to uphold its nonregulatory perspective on in vitro fertilization and assisted reproduction.

European Experiences

Like the United States, most European countries have taken a hands-off approach to regulating in vitro fertilization. A 1997 survey of fertility clinics in 30 countries found that "the majority of countries in Europe do not have established legislation pertaining to the various aspects of ART practice." According to researchers, only 10 of the 30 European nations had legislation in this area. Of those that did, most of the statutory language came from discussions organized by outside ethics committees or professional associations, rather than the government itself.[21]

This is consistent with the conclusion that much of state authority in complex policy areas has been delegated to experts and scientific societies. Biotech expert Jasanoff, for example, has characterized experts as the "fifth branch" of government. Through their work in advisory councils and professional associations, these "knowledge specialists" determine what practices are considered legitimate and who may have access to these particular services.[22] It is not government agencies, but scientific authorities who dictate much of public policy in the fertility area.

Assisted reproduction technologies vary considerably by nation. For example, about half of European countries allow these technologies for cohabiting couples, while one-quarter limit it to married couples, and another quarter allow single women to use the technology. Similarly, the age limit for donors varies from 30 years old in Israel to 55 years old in France. The storage time for frozen embryos ranges from one year in both Austria and Denmark to ten years in both Finland and Spain.[23]

In their overview of reproductive services in Europe, analysts Darren Langdridge and Eric Blyth classify country laws into categories of "laissez-faire, liberal, cautious regulatory, or prohibitive." According to their study, Greece, Hungary, and Poland are laissez-faire in their approach and place virtually no restrictions on use of assisted reproductive technology. Great Britain, the Netherlands, and Spain represent examples of a liberal, permissive approach to regulation. These nations have passed some legislation, but the laws do not dictate the content or form of treatment. Cautious regulators are Denmark, France, Norway, and Sweden.[24] Denmark restricts some practices, but does not restrict donor insemination and the country has wide public access to medical services. France, meanwhile, enacted legislation on the "Donation and Use of Elements and Products of the Human Body, Medically Assisted Procreation, and Prenatal Diagnosis" which allows assisted reproduction for therapeutic purposes only to "consenting infertile heterosexual couples."

At the other end of the spectrum, Germany and Austria have the strongest regulations on use of conception technologies. Mindful of its Nuremberg Code and Nazi medical abuses, Germany's Embryo Protection Act of 1990 limits reproductive therapies to in vitro fertilization and sperm donation.[25] The law does not allow egg or embryo insertion or surrogate motherhood. Through its 1992 Act on Procreative Medicine, Austria has these limits as well as restrictions on sperm donation.[26]

As a sign of the sympathetic public opinion environment surrounding assisted reproduction, 72 percent of Swiss voters in 2002 rejected a proposal to ban human reproduction "outside the body and [through] the use of donor sperm." Proponents of assisted reproduction technology called the vote a show of "solidarity with childless couples."[27] Had this nationwide referendum passed, the restriction would have been written into the Switzerland Constitution. An advocacy group collected over 100,000 signatures to place the item on the ballot.[28] But following a national campaign contested by both sides, citizens ultimately supported the right of patients to access fertility therapies.

In Great Britain, there was no government regulation of assisted reproduction technology from the first experiments in 1968 through 1985. Scientists and doctors were allowed to try different techniques as they saw fit. There were few formal restrictions on what they could do or how they employed new medical technology. In combination with commercial ventures, doctors and university professors were free to administer fertility treatments to childless couples.

Then from 1985 to 1990, medical officials practiced professional self-regulation through the Voluntary Licensing Authority established by the Medical Research Council and the Royal College of Obstetricians and Gynaecologists. This body had no legal authority to impose formal rules and requirements on physicians, but developed a code of practice. Although there were no mandatory regulations in place, most clinics observed the voluntary rules put in place by this authority.

It was not until the Human Fertility Act of 1990 that some formal government rules were established. The act mandates that physicians interested in practicing reproduction technology must apply for a license from the Human Fertilisation and Embryology Authority. That group's License Committee determines for whom and under what conditions medical professionals could make use of in vitro and other forms of reproduction technology. The committee does not stipulate particular acceptable medical treatments, but does make recommendations on "good practice."[29]

This regulatory approach has given doctors quite a wide range of latitude to employ reproductive technology and to incorporate new techniques into their practices. Regulation is limited to licensing rules and good conduct recommendations, and does not provide formal rules on how doctors may employ medical technology. Medical professionals have considerable leeway in exercising their own judgment and to market treatment options to the general public.

For those whose treatment is funded by the British National Health Service, who constitute about 25 percent of all assisted reproduction technology procedures, in vitro fertilization is limited to married, heterosexual couples. Single women and lesbians can seek treatment at private clinics, but not through facilities financed by the government.

In addition, women who treat themselves at home are unregulated because their actions are not considered a medical procedure. In a 2005 announcement of plans to rewrite the 1990 legislation, the government's public health minister Caroline Flint said, "We don't expect single women and lesbian women to be provided with treatment on the NHS. The current act talks of taking into account the welfare of the child and the need for a father."[30]

Data from 23 European countries in 2001 show that there were a total of 740 assisted reproduction technology clinics, out of which 579 clinics reported that they initiated 289,690 fertility procedures.

Germany had the highest number of interventions (72,000), followed by France (54,000) and Great Britain (35,000). These medical treatments produced a pregnancy rate of 25 percent per treatment.[31]

However, some forms of infertility treatment are banned in European countries. Oocyte donation has allowed women well beyond the traditional age range of 15 to 44 to become pregnant. With the prospect of women in their 50s and 60s becoming pregnant through female gamete donation, countries such as France banned egg donation for postmenopausal women. In practice, though, the enforcement of age-based restrictions in this kind of assisted reproductive technology has been difficult because of problems in determining the actual age of patients. In the last decade, women aged 63 in the United States, 60 in Great Britain, and 62 in Italy have given birth after deceiving fertility specialists regarding their real ages.[32]

Great Britain has placed legal restrictions on how long frozen embryos can be retained. Custody disputes between divorced couples and uncertainty regarding the length of time required for the preservation of married families' embryos led Great Britain to enact national legislation limiting the storage of frozen embryos to 5–10 years, depending on the circumstances. In contrast, the United States has left this decision to the states. The result has been great variation in state laws. Some states place no restriction on storage time for frozen embryos, while others require permanent storage.[33]

In 2006, the European Court of Human Rights handed down a landmark ruling saying divorced couples must agree when having children through embryos fertilized by a former partner. The decision came in a case where a British woman whose ovaries were removed due to cancer wanted to use an embryo to have a child. However, when her former fiancé refused to consent, she sued, but lost her right to use the embryo. This meant that both the man and the woman must be in agreement before embryos are used to bear children.[34]

Asian Experiences

Even more so than the United States and many European nations, Asian countries have adopted a permissive approach to assisted reproduction technology. Unlike Catholicism and some fundamentalist Protestant sects, Eastern religions, by and large, do not preach precepts that are interpreted as precluding assisted reproduction techniques. The objections that certain Western religions make about "playing God" or denying individual destiny are not a part of Eastern thinking. Many of these religions such as Buddhism, Shintoism, Confucianism, Taoism, and Hinduism are not rigid in their approach to everyday moral dilemmas.

Buddhists, for example, do not believe that using technology to assist procreation deprives individuals of "their right to an open future" or "their right to achieve individual enlightenment, or nirvana." They do not fear genetic determinism, and feel that individuals retain their basic humanity even if medical practices facilitate reproduction. Buddhists seek enlightenment and "freedom from bondage to the ego-self." Assisted reproductive technology is not seen as running contrary to meditation and the practice of "selfless behavior." Indeed, they are perfectly compatible with "ego-transcending conduct," the universal value of virtually all Buddhists. Only behaviors such as murder and theft are seen as obstructing the pursuit of nirvana.[35]

In the same vein, Hindu religion places few constraints on use of assisted reproductive technology. Its followers tell the story of "The Mahabharata," in which a pregnant mother divides a "mass of flesh" into 100 portions. Two years later, after storage in a pot and treatment with various herbs, the concoction turns into 100 Kaurava brothers. This precursor of modern-day in vitro fertilization is often cited in India as a justification for contemporary fertility therapies.[36]

Other Eastern religions have described manipulation of genetic material for reproduction, as "recycling," that is consistent with the belief of reincarnation.[37] Rather than viewing technology as a way for humans to play God, Asian scientists tend to interpret their religious beliefs as liberating them to pursue knowledge and innovation in all forms. They are unlikely to face active religious opposition in their own countries or to worry about the ethical overtones of assisted reproductive technology.

Public opinion about in vitro fertilization is positive in many Asian countries. A national public opinion survey in Japan, for example, found that people "see more benefits coming from science than harm when balanced against the risks." When asked specifically about in vitro fertilization, 58 percent described it as being worthwhile. The same survey in New Zealand revealed that 71 percent of the public thought in vitro fertilization was desirable. Another survey showed that around 60 percent of Indians, 50 percent of Japanese, 60 percent of New Zealanders, and 60 percent of Thais saw in vitro fertilization as a benefit to society.[38] A review of community surveys in Australia from 1981 to 2001 shows that support for in vitro fertilization has gone up to between 70 and 85 percent over this 20-year period.[39]

As a sign of its relative acceptance among Eastern religious leaders, a survey of Buddhist priests in Japan revealed that 43 percent approved of in vitro for married couples, while only 22 percent disapproved, and 35 percent were unsure.[40] While the support of in vitro among Buddhist priests is less than that of the general Japanese public, they still approved its use by a two-to-one ratio. This lack of religious opposition allows the scientific establishment to experiment in this area, relatively unconstrained by moral authorities.

Between the absence of opposition from organized religion and support from the general public, there are few legislative restrictions on the use of these technologies. In Japan, there is no legislation governing in vitro fertilization. The procedure is provided to married couples in private medical clinics, with oversight by a professional group, the Association of Obstetrics and Gynecology. Its guidelines do not permit surrogacy or oocyte donation, but do allow for sperm donation.[41]

China has an open policy on assisted reproduction technology as well as for the termination of both damaged or undamaged embryos. Since 1988, Chinese hospitals have allowed a variety of medical procedures designed to facilitate procreation. Techniques such as in vitro fertilization, gamete intrafallopian transfers, and embryo exchanges all are legal.[42] Although a 2001 Ministry of Public Health proposal required licenses to operate assisted reproduction clinics, a Chinese government review identified "44 sperm banks, 175 institutes that performed in vitro fertilization research, 126 institutions that carried out artificial insemination from unknown donors and 214 that used husband-artificial insemination had already been operating without government licenses."[43]

In addition, China passed a Maternal and Infant Health Care Law in 1995 that requires couples planning on marriage to obtain genetic screening tests. Anyone testing positive for genetic diseases, mental problems, or infectious illnesses must undergo sterilization or agree to use contraceptives to prevent pregnancy.[44]

As a society that remains authoritarian, China does not face much in the way of grassroots protests over biotechnology. It does not have a well-developed civil society and is able to allow genetic testing practices that would provoke moral outrage in other countries. Its mandatory screening law, for example, would not be tolerated in most Western democracies, but has not generated much popular opposition within China.

Although Hong Kong now is a part of China, it passed legislation in 2000 that licensed reproductive technology. No one was allowed to use this technology without having a license from the Council on Human Reproductive Technology. This council developed a code of practice that advised proper conduct in regard to assisted reproduction technology. The code allows for sperm and gamete donation, and embryo transfers. While the code was voluntary, reflecting Hong Kong's reliance on industry self-regulation, the council is allowed to take code violations into account when renewing licenses.[45]

India has no rules for licensing fertility clinics. In a society that places a high premium on childbearing, numerous clinics have arisen to assist couples.[46] A full range of assisted reproduction technology is offered in India, including in vitro fertilization, sperm and gamete donation, intra-cytoplasmic sperm injection, surrogacy, and embryo transfers. However, with no government regulations or licensing requirements, clinics vary considerably in their professionalism and honesty. Some service providers have been accused of inflating success rates, overcharging patients, or substituting less expensive procedures for more expensive ones. This has led to negative news coverage about industry practices.[47]

India operates a commercial system that is largely motivated by profit. Only those who can afford to pay for these services are able to access them. There are few incentives to standardize treatment. In a country where some people still consult faith healers in order to improve fertility, there are few restrictions on reproductive technologies other than those generated by the private marketplace.[48] India remains very permissive in its approach to infertility treatment.

Italian, Irish, and German Exceptions

While most countries around the world do not have strict regulations regarding in vitro fertilization, Italy, Ireland, and Germany are the exceptions. They have tougher rules on assisted reproductive technology than elsewhere. This reflects a particular combination of religious and cultural factors that allow the state to regulate procreation.

In Italy and Ireland, for example, the strong Catholic presence clearly contributes to the proregulatory impulse of these countries. With Italy being 96 percent Catholic and Ireland 92 percent, the church is a strong force in both places. Having an activist and politically powerful religious organization and a society with traditional attitudes has produced more restrictions than in other European countries.

Assisted reproduction technology in Italy is governed by a set of "circulars" and ordinances issued by the Ministry of Health. A 1985 rule known as the "Degan Circular" allowed married couples to request artificial insemination to overcome infertility, but prohibited the preservation of embryos for research and deferred implantation, and outlawed gamete donation. While sperm and gamete donation, and embryo storage for infertility were prohibited in National Health Services centers, such techniques were allowed in private medical clinics and were paid for entirely by the patient. A subsequent 1997 ordinance prohibited payment for gamete or embryo donation and required both public and private centers to report their activities in fertility treatment to the Ministry of Health.[49]

While these rules for public clinics adhere to Vatican teachings more closely than in most countries, they are not as strict as the Catholic Church would like. A 1987 Vatican doctrine entitled *Donum Vitae* ("the Gift of Life") clearly condemns the practice of assisted reproductive technology. In this encyclical, the late Pope John Paul II described in vitro fertilization as "contrary to the human dignity proper to the embryo" and "contrary to the right of every person to be conceived and to be born within marriage and from marriage."[50] In particular, the edict prohibited the freezing of human embryos, attempts to influence genetic inheritance, and the practices of sex selection, artificial insemination, and surrogate motherhood.

Like other countries, Italy relies on self-regulation by physicians. In 1995, the National Council of the National Federation for the Orders of Doctors and Dentists added a section on assisted fertility to its Code of Medical Ethics. According to the section's guidelines, doctors were forbidden to use artificial insemination except for stable heterosexual couples and widows. In addition, both surrogate motherhood and the advertising of gametes or embryos services on the part of fertility clinics were disallowed.[51]

Another code of self-regulation was issued in 1998 by a group representing researchers and private clinics. It established "minimum requirements for ART centres" and outlined exactly what kind of structures, personnel, and equipment were necessary for assisted reproduction. Although it was a voluntary guideline, most medical facilities follow its prescriptions. Doctors who object to reproductive technologies on grounds of conscience are not required to perform these treatments.[52]

By 2004, though, Italy enacted a tough new law regulating medically assisted reproduction. The legislation allowed a few infertility treatments, such as egg fertilization. For those techniques that were legalized, the law imposed strict procedures that doctors must employ when treating couples. However, it outlawed techniques that are commonly used in other European countries, such as third-party sperm or oocyte donation. For example, Great Britain, France, Greece, and Spain allow for sperm and oocyte gifts, while the Italian legislation does not. In addition, surrogate motherhood was prohibited, as was the freezing of embryos for purposes of reproduction. The number of eggs that could be fertilized was limited to three, and all must be used at the time of treatment. None are allowed to be saved for use at a later time. Sharp limits were imposed on preimplantation testing for genetic disorders. All embryos were defined as full human beings and were required to be implanted, regardless of whether there were genetic disorders.[53]

According to medical observers, Italy's new law is the most stringent in Europe. It had an immediate effect on fertility treatment in that country. The techniques allowed in Italy produce "less than half as many pregnancies as the usual care does in many cases.". The result was that those who could afford to travel to other countries to get better treatment did so.[54] The availability of fertility treatment in other European Union nations makes it possible for those who can afford it to avail themselves of assisted reproduction.

Unhappy over this outcome, secular forces pushed for a national referendum designed to weaken the 2004 legislation. In June 2005, Italians went to the polls to express their view of the new law. But the referendum failed after Catholic leaders encouraged voters to stay home, and only 26 percent of Italians voted. A constitutional requirement that at least 50 percent of the electorate had to cast ballots for the referendum to be legitimate meant that the tough new rules put into effect through the 2004 law were left intact. It was a strong show of support for the Catholic Church's restrictive views toward assisted reproductive technology.[55]

Ireland has relatively few in vitro procedures (around 4,000 per year), compared to other European nations. At the country's nine fertility centers, there are waiting lists of at least four years. The Irish government does not provide any financial support for couples undergoing the therapy, unlike most other European Union countries. For example, the National Health Service in Great Britain pays for one free in vitro treatment for females under 40 years of age. Other European nations provide the same type of financial support.[56]

Until recently, there had been little governmental regulation of assisted reproduction technology in Ireland. Eight of the private clinics offer sperm freezing, six offer in vitro fertilization, four provide sperm donation, four offer artificial insemination from the husband, and one offers ovum donation. A few provide for embryo storage. Physicians employing these techniques are subject to guidelines developed by the private Medical Council, which is Ireland's trade association for doctors. However, nurses and scientists are not subject to the council's code of ethics.[57]

This would be changed, however, by legislation proposed by the Commission on Assisted Human Reproduction in 2005. It would regulate all aspects of infertility treatments. Clinics providing this kind of therapy would be required to register with the commission and medical personnel providing infertility assistance would be subject to strict rules. Surrogate parenting would be outlawed and donors would not be allowed to be paid for sperm or egg donations.[58]

The limited availability and long waiting period for assisted reproduction in Ireland has encouraged couples to travel to Spain and other Eastern European nations, where procedures are more readily available. This new industry known as "fertility tourism" offers low-cost treatment with short waiting periods. Hungary and Slovenia, in particular, have carved out a niche for themselves in fertility treatment. For example, Slovenia has provided 1,222 treatments per million people, compared to 963 in the Netherlands and 593 in Great Britain. Costs of these procedures generally are lower in Eastern than Western Europe.[59]

Some Irish couples have even reported going abroad to the United States, where laws allow surrogate mothers and provide for preimplantation genetic testing to avoid the conception of children with hereditary disorders.[60] In an area with strong demand and mobile patients, the existence of a global market makes it difficult for Ireland to enforce its restrictions on those who can afford to seek treatment elsewhere.

One of the reasons Ireland has been slower than Italy to regulate assisted reproduction technology is that its Catholic residents do not totally share the church's tough stance on this issue. Despite Ireland's predominantly Catholic population, two-thirds of its citizens favor in vitro fertilization for married, heterosexual couples. According to a national survey undertaken by the country's Commission on Assisted Human Reproduction, small majorities also were in favor of the use of surrogate mothers for infertile couples and donors providing sperm or eggs in these treatments. The country was much more divided on whether such treatments should be provided to same-sex couples, single women, or older women.[61] In contrast to Italy, Ireland's favorable public opinion towards in vitro fertilization counterbalances the ability of the Catholic Church to use the state to regulate reproduction.

Meanwhile, in the aftermath of World War II, Germany enacted tough ethics rules on medical research. Nazi experiments on nonconsenting patients led to national regulations regarding use of medical procedures. Contemporary political pressure from the Green party, liberals, and feminists reinforces this concern. Rather than trusting the private sector or industry self-regulation, the state took on the role of overseeing fertility issues.

In the area of assisted reproduction technology, Germany imposed a number of restrictions on both public and private clinics after assisted reproduction technology became available in the 1980s. Based on the 1990 Embryo Protection law, gamete donations are prohibited, as is surrogate motherhood, embryo research, and preimplantation genetic testing. These are practices that are allowed in most other European countries (with the exception of Ireland and Italy), so clearly German history has colored the utilization of these kinds of biotechnology in that country.

In addition, posthumous in vitro fertilization and embryo freezing are outlawed. Embryos are protected by Germany's Basic Law and given the same rights all humans have. Transferring more than three embryos to the womb is defined as a crime and is punishable by up to three years in prison. It is illegal to fertilize more eggs than can be implanted in a woman during a single cycle.[62]

The result of these tough rules is that German success rates are much lower than in the United States. While the American pregnancy level is 37.8 percent per in vitro transfer, it is 27 percent in Germany. Similar to the situation in Italy and Ireland, some German women who seek assistance with infertility problems are forced to travel to other countries to avoid restrictions in their own country.

Conclusion

This cross-national review of assisted reproduction regulation in various countries demonstrates the difficulties of regulating procreation in a globalized world. In most places, public opinion is decidedly favorable toward infertility treatment. Governments have been more likely to rely on industry self-regulation and professional guidelines than to require overt regulation. Patients can shop from country to country for rules that are the least restrictive. As long as patients can pay for the treatment, geographic mobility allows them to bypass regulations in particular nations.

These factors undermine the ability of nation-states to restrict access to new technologies that are seen as desirable. While there are exceptions to this general tendency in the cases of Italy, Ireland, and Germany, most countries essentially have privatized decision making in this area. Commercial enterprises, doctors, scientists, and patients are trusted to make important decisions. With the high cost of in vitro techniques, this science-industrial complex has proven quite lucrative. Between country-shopping by patients and collaboration between industry and science, the ability of governments to regulate the marketplace is quite limited.

In the national exceptions, religion and culture clearly play a major role in allowing governments to impose stronger restrictions on fertility treatments. The strong Catholic presence in Italy and Ireland and the residue of Nazi experiments in Germany have enabled those nations to enact tougher laws regulating assisted reproduction. This demonstrates that the science-industrial complex is susceptible to regulation where religion and culture are particularly strong.

Chapter 4

Genetically Modified Food

Genetically modified foods have been a major growth area in biotechnology. In 2005, 21 nations raised biotech crops on a commercial basis, up from 17 the previous year. Around 90 percent of Australian cotton and 80 percent of American soybeans currently are based on altered crops. The countries producing the most engineered products are the United States, Argentina, Canada, Brazil, China, Paraguay, India, South Africa, Uruguay, Australia, Romania, Mexico, Spain, and the Philippines. Together in 2005, these countries planted 222 million acres of genetically modified crops, up from 167 million acres in 2003. These numbers represent a dramatic increase over the 4.3 million acres planted in 1996, the first year genetically modified foods were commercially produced.[1]

In looking at how various countries handle genetically altered foods, there is tremendous variation in the level of regulation.[2] Similar to in vitro fertilization, the United States has adopted a laissez-faire stance on biocrops, but not meats, while a number of European and some Asian nations regulate and discourage all kinds of engineered foods. The American approach stems from the strength of agribusiness lobbying and a desire to export food to other countries. In contrast, many other nations around the world have been much more cautious about this particular biotech area. They worry about the safety of the food supply and hidden health risks associated with genetic modifications. International environmental groups have lobbied successfully for labeling and disclosure requirements, and for limits on field experimentation. Health and safety concerns among the population as a whole have led these nations to regulate food engineering, require food labels for consumers, and demand more detailed oversight of commercial activities.

Overview of Genetic Modifications

According to experts, genetic modifications refer to "altering the genetic trait of a living organism by using recombinant deoxyribonucleic acid (DNA) technology to transfer one or more genes from one organism to another by crossing biological species."[3] New advances in molecular and cell biology allow scientists to modify foods by the identification, copying, and insertion of genes into plants. Using a process known as "transformation," scientists add genes with useful qualities into a plant's basic structure and reproduce a new structure with those traits. This gives crops the ability to have much greater shelf life and resistance to diseases and pests.[4]

In the past, it was possible to develop new varieties only within the same species. It was a laborious process to generate new hybrids of basic crops such as wheat or corn.[5] Scientists had to cross-pollinate different crop varieties, see what desirable traits resulted, and reproduce those hybrids. Such efforts took years of trial and error research to produce new varieties that had the qualities producers desired.

With the development of recombinant DNA technology, genetic engineering techniques have speeded up the process of creating new varieties and expanded hybridization beyond individual species.[6] Rather than having to create new crops, plant them, and see what emerged, scientists now can modify the gene structure of particular foods and select traits that are considered virtuous. Through gene manipulation, "green" biotechnicians have developed food crops with greater yields, lower requirements for pesticide use, and increased resistance to particular diseases and pests.

The result has been an explosion of genetically altered foods. About 10 percent of the world's seed supply now includes biocrops.[7] Although biocrops are generated by a few large countries (such as the United States, Argentina, Canada, Brazil, and China), global trade has spread engineered food around the world. In some countries, more than two-thirds of the seeds for major food crops are based on genetically modified organisms.

The "Big Three" bio-crop companies of Monsanto, Syngenta, and Aventis currently control about half of the world's agribusiness.[8] Mergers and acquisitions have turned these firms into multi-billion-dollar enterprises with tremendous political power.[9] Monsanto, for example, has 9 full-time lobbyists on its staff plus another 13 who work for them at private firms. Between 1999 and 2004, the company spent $18.5 million on lobbying. Monsanto's prowess allowed it to block labeling requirements in the United States, to keep Congress from holding even a single hearing on the subject of genetically modified food, and to press the European Commission and Asian nations for relief from national bans on genetic engineering.[10]

These and other companies have developed partnerships with scientists in many countries. According to Monsanto officials, the company finances bio-crop experiments around the world and relies on scientists from universities and research institutes to develop new food varieties through recombinant DNA.[11] As discussed in chapter 2, these partnerships represent revealing examples of "country-shopping" and "scientist-buying" that are the hallmarks of biotechnology. If countries such as Germany or Japan crack down on altered foods, agribusinesses merely shift their efforts to other nations that have fewer qualms about and restrictions on genetically altered food. And like the earlier example of China and genetically modified rice, in cases where the government has specific regulations, contraband trade brings illegal seeds despite the formal prohibition on engineered crops.

In spite of the political power of agribusiness, there is considerable worry in many places that genetically modified foods are not as safe as those found in nature. Critics question the adequacy of current testing procedures to determine the health and safety of food products. There is a sense in some circles that new, engineered products are being created without sufficient safeguards of the public's health and well-being. These concerns have led to wide variation in how different countries handle engineered food products.

The American Experience

In the United States, there is little government regulation of genetically modified plant foods, but restrictions in regard to meat do exist. In regard to the latter, scientists earned national press attention when they announced they had cloned a pig that was able to generate its own omega-3 fatty acids. These materials are the acids found in oily fishes that are associated with reduced risk from heart disease. As genetically modified meat is still in the experimental phase, the U.S. Food and Drug Administration(FDA) has not approved any altered meat for human consumption. Consumer groups said if companies attempted to gain government approval, they would fight the effort. Joseph Mendelson of the Center for Food Safety, a nonprofit group opposed to genetically modified food, said consumers would not tolerate this kind of food. "I am confident that consumers would not want them," he said.[12]

The situation with plants, however, is far more lenient. Similar to the case of in vitro fertilization, the public sector has allowed agribusiness to plant and market products with little government regulation. Industry groups such as the National Farm Bureau Federation, National Corn Growers Association, and the American Soybean Association are very sympathetic to biocrops and have been successful at lobbying government to avoid strong regulation.[13]

Most of the responsibility for overseeing food safety rests with the FDA and, in the case of field trials, with the U.S. Department of Agriculture. In 1986, the FDA adopted its first policy known as the Coordinated Framework for Regulation of Biotechnology. This directive gave it the authority to regulate recombinant DNA and bioengineered food products. Under legislative power authorizing the FDA to secure the safety of the nation's food supply, the framework specified that bioengineered crops must meet the same safety standard as other food products. If safety problems arose, the FDA could force particular food items to be withdrawn from the market until concerns were met.[14]

In 1992, the FDA altered its previous policy by adopting a new statement on food and animal feeds. In particular, it extended government oversight to the "pre-market" stage of food production and added regulatory oversight to new varieties of plants that were commercially sold. This allowed it to examine whole foods, including products developed from grains, vegetables, fruits, food starch, or vegetable oils. Previously, these items were not seen as requiring pre-market FDA approval.[15]

Among the first products approved were the enzyme chymosin (rennet), for use in cheese and dairy products, and Calgene's "Flavr Savr" tomato. According to an FDA report, these food products demonstrated "improved shelf-life, processing characteristics, flavor, nutritional properties, and agronomic characteristics, such as tolerance to chemical herbicides and resistance to pests and diseases." Chymosin was judged to be "generally recognized as safe." This standard allowed the FDA to exempt the new creation from pre-market approval required of food additives.

In its justification of this decision, FDA officials said that "the introduced chymosin gene encoded a protein that had the same structure and function as animal-derived chymosin; the manufacturing process removes most impurities; the production microorganisms are destroyed or removed during processing and are non-toxigenic and non-pathogenic, and any antibiotic-resistance market genes (e.g., ampicillin) are destroyed in the manufacturing process."[16]

In the case of the Flavr Savr tomato, the commercial company Calgene added an antisense polygalacturonase gene to the product. The new gene suppresses the enzyme associated with the breakdown of pectin in the tomato skin, and allows farmers to keep the tomato on the vine longer before shipping to market. According to the company, the new genetically altered tomato would ship better and be more flavorful for consumers. In testing undertaken by Calgene, scientists compared the new tomato with its parent line of ordinary tomatoes, and claimed the products to be roughly equivalent.

This precedent allowed the FDA to develop the doctrine of "substantial equivalence" for genetically altered food. According to this standard, if an engineered product is genetically similar to an existing "natural" crop, then the agency considers it to be a natural product for all practical purposes. The new item can be sold, marketed, and consumed without fear of negative impact on health or safety.

With the Flavr Savr tomato, there was some concern regarding impact on antibiotics usage. Calgene examined the effect of its gene marker on the efficacy of antibiotics and demonstrated "that only a small fraction of the antibiotic would be inactivated." Based on the company's studies and the FDA's scientific review, the government judged the Flavr Savr to be safe for consumers.

Between the genetic testing and the rule of "substantial equivalence," the FDA decided not to require any special labeling to notify consumers of the tomato's genetic alterations. Genetically altered foods were safe and consumers did not need to worry about danger to their food supply. This decision stands in stark contrast to policies requiring consumer labeling in other countries. In most European nations, for example, governments decided consumers had the right to know what they were eating. Detailed labeling instructions informed people of the difference between "natural" and "altered" food, and consumers would decide the health risks through their own consumer choices. In contrast, the FDA required special labeling only if the genetically modified food "differs significantly from its conventional counterpart."[17]

Within a few years, the FDA had approved a number of genetically modified foods. This included delayed-ripening tomatoes (DNA Plant Technology, Monsanto Co., and Zeneca Plant Sciences); pest-resistant crops and virus-resistant squash (Asgrow Seed Co.); Colorado potato beetle-resistant potato (Monsanto Co.); herbicide-tolerant crops such as bromoxynil-tolerant cotton (Calgene Inc.); and glyphosate-tolerant soybean (Monsanto Co.).[18]

In reviewing government actions since that time, official watchdogs have found that much of the U.S. Department of Agriculture's oversight of field trials in the United States has been lax and ineffective. A 2005 report by the agency's own auditor, for example, concluded that "biotechnology regulators did not always notice violations of their own rules, did not inspect planting sites when they should have and did not assure that the genetically engineered crops were destroyed when the field trial was done." These conclusions were based on visits to 91 field sites between 2003 and 2005.[19]

The general weakness of American government regulation helps to explain one of the worst instances of food contamination that has occurred in the United States. In 1998, a commercial company named Aventis CropScience developed a new variety of corn called "StarLink." It carried a *Bacillus thuringiensis* (Bt) toxin that was effective at repelling major insects.

Based on its potential as a replacement for popular insecticides, the U.S. Environmental Protection Agency approved the corn for use in cattle feed. By 2000, though, traces of StarLink corn were found in corn tortillas and other processed foods. How this genetically modified substance moved uninterrupted from animal to human food became a source of great controversy.[20] Because there is an alleged link between the Bt toxin and food allergies, critics cite this "escape" as a sign of the impossibility of keeping animal food separate from that intended for human beings and the laxity of government oversight of genetically modified products.

One of the reasons for the laissez-faire approach to food safety in the United States has been the power of agribusiness political lobbying. Industry groups have lobbied Congress not to place very many restrictions on the development of genetically modified foods. Unlike many European and Asian nations, these associations have been successful at avoiding labeling on Americans foods to let consumers know that some of the items in their grocery store are the result of genetic alterations.

Commercial companies have argued that genetically modified foods are a better product and help the United States ship food to the rest of the world. Indeed, the export of genetically altered foods has been a major growth area for U.S. agriculture. With so many countries needing to import food, American companies have emphasized the economic payoff of this new biotechnology.

Another cause of the weakness of U.S. regulation in this area is general public unawareness, as evidenced by public opinion surveys showing that most Americans know little about genetically modified foods. In 2003, the Pew Initiative on Food and Biotechnology commissioned a national poll of 1,000 consumers. The results found that only 34 percent of Americans said they had heard a great deal or some information about genetically modified foods. This was down 10 percentage points from a similar survey undertaken in 2001.

Over the last few years, public opposition has dropped. Whereas 58 percent of Americans in 2001 opposed the introduction of genetically modified foods into the U.S. food supply (and 26 percent favored it), the number against doing it in 2003 declined to 48 percent (with support coming in at about the same level of 25 percent). When asked if they would eat genetically modified food, 43 percent said they would (up from 38 percent in 2001), while 50 percent indicated they would not (down from 54 percent in 2001).[21]

A 2001 survey undertaken by the Food Policy Institute at Rutgers University found that 58 percent of Americans thought they had never eaten any genetically modified food and half were unaware that genetically altered food currently was even sold in the United States.[22] These results are striking because experts estimate that 60 to 70 percent of American processed foods include genetically modified ingredients. According to the Food Policy Institute, "A majority of the soy and rapeseed (canola), and a third of the corn harvested in the United States and Canada in 2002 were GM varieties."[23]

Although Americans feel they have not been exposed to altered crops, the reality is that they have. With the large amount of genetically modified food in American grocery stores, the odds are that most people have consumed this kind of produce. Farmers in the United States are growing a high percentage of genetically modified food products, especially soybeans and corn, and these are staples of the American food supply.

In general, Americans oppose a ban on genetically modified food, but want companies to submit detailed safety information to the FDA. Ninety-four percent think altered food should be labeled as such.[24] Eighty-nine percent of Americans think that food companies should make these kinds of submissions to the government. When asked on a 0 to 10 scale how comfortable they are with genetic modification of various kinds of life forms, Americans were most comfortable with changes in plants (mean of 6.08), compared to microbes (4.24 mean), animals for food sources (3.81 mean), insects (mean of 3.61), animals for other purposes (2.27 mean), and humans (a mean of 1.31).[25] But despite these sentiments, the United States has adopted no labeling requirements on engineered foods.

European Experiences

Unlike the United States, most European countries plus the European Union have adopted a more restrictive stance on genetic food modifications. Because of their tighter restrictions, the result is there are only 247,000 acres of modified foods in production for the continent as a whole, well below the level of the United States, Canada, Argentina, Brazil, and China.[26] National bans on genetically altered food are in force in five European countries: Germany,

France, Austria, Greece, and Luxembourg. De facto moratoriums on the commercial sale of genetically modified organisms are in place in Denmark, Italy, and Belgium.

In 1997, the European Commission issued its Regulation on Novel Foods and Novel Food Ingredients. It established strict rules for the labeling and sale of genetically altered food. This was followed in 2001 with a Directive on Deliberate Release of Genetically Modified Organisms (GMOs), which created a process for deciding whether altered foods could be released into the environment.[27] Between 1998 and 2004, the European Commission banned the purchase of genetically modified organisms shipped from the United States.

This led American leaders in 2003 to file a complaint with the World Trade Organization that accused Europe of illegally restricting food imports. According to the complaint, the European Commission's moratorium on genetically modified foods "violated a global treaty on standards for food, which requires governments to act without 'undue delay' and to base decisions on scientific risk assessments, not political expediency."[28] The World Trade Organization upheld this complaint and ordered the European Commission to open its borders for genetically modified food.[29]

In 2005, the Commission lifted its ban on canola in animal food products and allowed certain engineered seeds to be sold in countries without formal prohibitions. But a number of European nations retain their ban on engineered food. The result has been little production or marketing of altered foods in Europe.[30] In an effort to force its will on member nations, the Commission in 2006 ordered Greece to remove its national ban on genetically modified corn. Even though Greek farmers strongly opposed the policy change, the body said that in the absence of evidence showing health or safety problems, Greece must open its agriculture to these products.[31]

Most European nations and the European Commission have stiff labeling requirements in place for genetically modified foods. The Commission has considered a proposal that would allow up to three genetically modified seeds in batches of 1,000 conventional seeds without informing consumers or food manufacturers. However, when 10 of the 25 commissioners opposed this change in labeling procedures and 5 were undecided, the Commission upheld their previous stance.[32]

Among European countries having food labeling requirements, there remains uncertainty over how to define the term "GM-free," and exactly what information to provide consumers. Austria has the clearest stance as it has set the "level of permissible contamination of protein encoded by foreign genes at 0.1 percent of product weight." Germany has had political battles over "low-level" contamination involving genetically modified enzymes. One cartoonist joked that the new German compromise label should read: "GMO-free unless accidentally contaminated, these things happen from time to time, you know, even in Bavaria."[33]

The American government has complained about the European Commission's hostility to engineered plants and the resulting damage to commercial trade. Whereas U.S. corn exports to Europe totaled 3.3 million tons in 1995, the figure dropped to 25,000 tons by 2002 as a result of Europe's antipathy to biocrops. This drop deprived U.S. farmers of around $300 million in corn sales each year. American representatives lobbied the World Trade Organization that, in the absence of scientific studies showing a threat to food safety, European bans represented a protectionist measure that was forbidden by global trade agreements.[34]

There is variation within Europe in how different countries have responded to altered food. While France, Germany, and Austria remain the most hostile to genetically engineered crops, Spain, Portugal, the Netherlands, and Great Britain are more sympathetic. In Spain, for example, genetically modified foods are expanding considerably. Whereas there were 32,000 hectares (or 79,073 acres) of altered crops in 2003, the amount had grown to 58,219 hectares (or 143,862 acres) in 2004, an increase of 82 percent. In order to keep modified crops from contaminating natural farming, Spain has decreed a minimum distance of 25 meters (or 27 yards) between the two fields.[35]

Great Britain has relied on decentralized regulation. Rather than having a single agency that monitors genetically altered food, such as the FDA in the United States, biotech firms wanting to grow engineered crops must deal with a variety of government departments. The resulting patchwork of oversight has led to little continuous monitoring and concern on the part of experts that more oversight is necessary. The British Royal Society has called for a single monitoring agency that would regulate all activities in this area and end practices such as the "use of antibiotic-resistance marker genes in genetically modified food products." However, the society stopped short of adopting stronger regulations similar to those in other European nations. British government agencies did not favor mandatory labeling requirements for genetically engineered food and indicated they were not worried about spreading genes from altered plants to naturally occurring ones.[36] In 2004, these recommendations were implemented in Great Britain and periodic inspections were mandated to insure that oversight rules were being followed.

Even in places where genetically modified crops are forbidden by law, authorities acknowledge contraband trade. For example, Turkey forbids use of genetically modified seeds. Yet its officials concede that "because of the poor level of controls at Turkish customs we suspect that corn, cotton and tomato seed is getting into Turkish fields."[37] The same thing happened with engineered corn produced by Syngenta, a Swiss company. Reminiscent of the infamous StarLink "escape" in the United States, the company mixed corn with Bt-11, which is legal in Europe, with Bt-10, a genetically altered strain that has not been approved in Europe. The latter has a gene that is resistant to ampicillin, a popular antibiotic.[38] These episodes reveal the difficulties of enforcing formal rules when borders are porous, such as is the case in the European Union.

One of the explanations for the tough European stance on genetically modified organisms is public opposition. Thomas Bernauer and Erika Meins have pointed out that many European countries do not have a strong biotech sector so demands for freedom from regulation are relatively weak compared to the United States. At the same time, opposition groups and nongovernmental organizations (such as Greenpeace, Friends of the Earth Europe, and European Bureau of Consumers' Union) are strong, and there is an institutional framework that allows these forces to express their concerns about genetic engineering.[39]

Public opinion data from European locales show greater fear about genetically modified food than is the case in the United States. According to one set of researchers, "People in the United States were significantly more supportive of GM crops and GM foods than were people in Europe."[40] European consumers worry that untested combinations pose a safety hazard that scientific studies are unable to detect. They also have less trust in the ability of their nation's regulators to be effective, compared to the more favorable views of Americans about U.S. regulators.[41] These views have marked the outcome of referendum campaigns on genetically modified food. In 2005, for example, 55.7 percent of Switzerland voters approved a five-year ban on the production of biocrops.

In Great Britain, a survey found that 77 percent of respondents said they wanted genetically modified crops banned and 61 percent indicated they prefer not to eat such foods.[42] Another study demonstrated that British consumers were willing to pay more money in order to avoid genetically engineered foods.[43] Only 13 percent of Italians said they would purchase genetically altered food, even if it were discounted to the consumer.[44]

Institutional barriers in Europe have slowed innovation in green biotechnology. Under current rules for European Commission approval of new regulations, proposed changes require two-thirds of European Union states to adopt the proposals. In a contentious policy area with strong opposition groups and unfavorable public opinion, this has slowed decision making to a standstill. Even when the Commission finds certain altered crops to be safe, those recommendations almost never meet the two-thirds majority rule for adoption across the European Union. The result has been difficulty in changing European decision making toward a more sympathetic view of genetically modified food.[45]

Asian Experiences

Food production is a particular problem in many Asian nations due to the large populations of many of its countries. Of the world's 6.5 billion people, half (3.4 billion) live in Asia. China has around 1.3 billion residents, while India has 1 billion, Indonesia 212 million, Pakistan 135 million, Bangladesh 128 million, Japan 126 million, and the rest reside elsewhere in Asia.[46] It is estimated that 900 million Asians are poor, based on international poverty statistics, and 536 million are undernourished.[47]

The combination of substantial populations, extensive poverty, and limited areas for agriculture make many Asian countries particularly open to new food production techniques. Agriculture accounts for 20 percent of Asia's overall GDP and employs half the workforce.[48] In the

1970s and 1980s, Asian governments supported the Green Revolution that brought crop hybrids and fertilizer usage to that region's agricultural practices. They devoted substantial resources to new rice hybrids, increased use of fertilizer and pesticides, and greater employment of irrigation techniques. Between 1970 and 1995, the amount of cereal crops raised in Asian nations doubled and there was a significant increase in caloric food availability to the people.[49]

With the advent of genetically modified foods, some Asian countries have expressed great hope in the ability of this biotechnology to improve food production. In particular, China, India, Indonesia, Malaysia, the Philippines, and Thailand are incorporating engineered foods into their economies. They are spending considerable money on new products, and encouraging their companies to make use of new technologies.

The Philippines was the first Asian nation to allow commercial production of Bt corn. With assistance from Monsanto, the government authorized the planting of 20,000 hectares (or 49,420 acres) in 2002. This genetically modified corn uses Bt to resist corn borer pests. Through its Bureau of Plant Industry, commercial companies must get government approval for the sale of engineered biocrops, but the government has been supportive of several new varieties.[50]

China has made a big push on genetically modified crops. Indeed, it has in fact emerged as the world's second-largest investor in bio-crop research, after the United States. In 1999, it devoted $112 million to new research in this area. One particular reason China is interested in developing this kind of agriculture is that too many of its farmers are moving to urban areas and there is a need to improve the efficiency of crop production among the remaining workers in order to feed its people. "As more and more able-bodied farmers leave villages to seek better paid jobs in cities, women and old people are doing more of the work. GM rice can help alleviate their workload, and reduced pesticide use will improve their health and the environment," explains a Chinese biotech researcher.[51]

With planned spending increases, China will be responsible for about a third of the world's expenditures on plant biotechnology. Thirty-one genetically altered crops have been approved for commercialization there, and another 100 are in the testing stage. Among the crops approved are Bt cotton, sweet peppers, tomatoes, and petunias. In 2004, Chinese farmers raised 3.7 million hectares (or 9.1 million acres) of genetically modified cotton. This represented a 32 percent increase over 2003.[52] Bt cotton now constitutes 65 percent of the country's entire cotton crop.[53]

The country's National Biotechnology Development Policy Outline was developed in 1986 to guide biotech policy. It represented a collaborative effort between 200 scientists and government officials directed by the State Development and Planning Commission, the Ministry of Sciences and Technology, Ministry of Agriculture, and the State Economic Commission. An Agricultural Science and Technology Development Compendium was announced in 2001 with the goal of improving agricultural productivity.

Befitting a command economy, much of China's investment in agricultural biotechnology comes from the public sector. China has 150 labs devoted to plant and animal biotechnology, and commercial donors provide just 6 percent of their research budgets.[54] However, more and more collaboration is taking place between research institutes and commercial companies. Monsanto has worked with scientists on engineered maize, cotton, and rice; while Syngenta is helping to develop genetically modified rice; and Pineland and Delta have cooperated on Bt cotton. Through these science and industry collaborations, the hope is that China will leap forward and use biotechnology to feed its large population.

In a development that shows how quickly China is shifting toward capitalism, some scientists are starting to establish commercial companies in order to sell their expertise. These firms are marketing biotech products and helping companies commercialize new knowledge in this area. Much like their American counterparts, Chinese scientists are marketing their skills locally as well as to multinational corporations. Their companies are being listed on Chinese stock exchanges.[55]

For products approved for commercial release, China's Ministry of Agriculture oversees plant experimentation.[56] In one set of field trials, scientists at the Institute of Genetics and Biology of the Chinese Academy of Sciences found that an altered rice crop was especially resistant to pests such as moths and butterflies, two of the major pests that are harmful to rice. The Ministry has an Office of Biosafety that judges the safety of engineered crops. In one safety study, scientists fed altered rice to rats for several generations. After seeing no ill effects in the rats, they concluded the rice was safe.[57]

In the face of international concern about genetically modified foods, China has started to exercise more serious oversight. In 2002, the country started labeling genetically altered food products so consumers would know what was being purchased. However, its standards for labeling rules are more lenient than what is common practice in many European nations.

In 2005, China ratified the Cartagena Protocol on Biosafety (unlike the United States, which has not yet endorsed this protocol). This is an international agreement that governs trade in genetically modified organisms. Part of the Convention on Biological Diversity, it became effective in 2003 and has been signed by 120 countries.[58] Among its provisions are requirements that exporters have the consent of importers before shipping genetically modified foods, detailed documentation of trade commodities, and reporting new legislation or regulations to a Biosafety Clearinghouse within 15 days of enacting the rules.[59]

Although sale of engineered rice is currently illegal in China, Chinese researchers are working on modified rice that is resistant to major pests.[60] China has been cautious about introducing altered rice because it is a staple of the food supply, and unlike cotton for example, it has several wild varieties. This increases the risk that plant traits will spread quickly from cultivated to naturally occurring varieties. Such fears may in fact be justified. When Bt cotton was introduced in 1998 in just two Chinese provinces, critics later claimed that it spread rapidly beyond the control of government agencies. According to Sze Pang Cheung of Greenpeace in Beijing, a group opposed to genetically modified foods, "there is no way to block it. The farmers exchange the seed, they carry it to other places. The seed companies set up shops with people who don't even know about the regulations."[61]

As a sign of the difficulty government officials have in controlling contraband trade, transgenic rice has been found in markets located in Hubei Provence in the middle of China. The rice bags bore the label of the Huazhong Agriculture University in Wuhan, where scientists have been experimenting with the new rice varieties. Even though it is illegal to sell genetically modified rice, farmers in that area are using what they call "anti-pest rice." One farmer interviewed said, "This is really remarkable rice. All you do is plant it and it grows. You don't need to use all those chemicals any more."[62]

A scientific study coauthored by Chinese and American scientists found that "genetically modified rice cut pesticide use by as much as 80 percent." That project argued that if 90 percent of Chinese farmers switched to engineered rice, national agricultural income in China would rise by $4 billion.[63] The laxity of regulatory approval for the sale of genetically modified rice may make China the first country to allow a major altered food staple to be directly eaten by humans without government approval.[64]

A 2003 Greenpeace survey of residents in Guangzhou, China found that "most of the 1,000 consumers polled preferred food commodities containing no GM ingredients, while 87 percent thought transgenic food products should be labeled." In addition, 50 percent claimed they would be willing to pay more money for food products that were not genetically engineered.[65] But despite evident citizen concern, China continues to move ahead with field trials, and more and more of its agriculture is based on engineered crops.

India is another biotech leader in the region. In 1982, it established a Biotechnology Board in the Department of Science and Technology. This was followed by the 1986 creation of the Department of Biotechnology. That administrative unit oversees the testing and safety of transgenic crops in India.[66]

Much of Indian activity in biotech has been based on public-private partnerships. Unlike China, much of the money devoted to biotech research and development comes from the private sector. Monsanto, for example, has a subsidiary called the Maharashtra Hybrid Seeds Company, the largest private seed company in India. The joint venture has invested $100 million in the development of Indian transgenic crops, and works with Indian research institutes to develop new biocrops and to test engineered food products.[67]

India has 50 research institutes devoted to plant biotechnology. Of the $25 million spent in this area, $15 million comes from the public sector while $10 million is raised from the private sector. In 2002, India released its first genetically modified crop, Bt cotton, following approval of the government's Genetic Engineering Approval Committee.[68] The country's high interest in genetically modified cotton is spurred by the fact that 30 percent of the nation's cotton production is damaged by disease and pests.[69]

Similar to the United States, India does not have a labeling requirement for genetically altered food.[70] India has not followed the lead of European countries and mandated food labels. Instead, it has trusted the private sector to sell safe food products and does not feel the need to warn consumers about genetically modified food.

This does not mean, however, that there is no opposition to transgenic crops. Farmers in some states have burned field experiments due to fear of contamination. The government in Andhra Pradesh was forced to terminate a field trial due to local opposition. But despite these protests, the government places a high priority on biocrops because of the "urgency to enhance food production and develop crops with desired traits." One official in the Department of Biotechnology stated that "with 60 million acres of transgenic plants under cultivation worldwide, India cannot lag behind others in this technology."[71]

Other Asian countries, though, have not been as receptive to transgenic crops. Japan, for example, is not as enamored with genetically modified food. In 2001, it adopted a labeling requirement for products containing more than 5 percent genetic modifications that is a lower threshold than the 0.9 percent standard popular in many European countries, and the 3 percent threshold in South Korea.[72] This was part of new national regulations issued by the Ministry of Agriculture, Forestry, and Fisheries. Labels are required for whole foods and processed food products in Japan.[73]

As the world's largest importer of food, Japan has been very careful about safeguarding its food supply. After cases of mad cow disease, mislabeling of beef, and genetically modified corn that was surreptitiously introduced into the food chain, Japan froze genetically altered rice trials in 2004.[74] There also was a public outcry when Syngenta shipped genetically modified corn to Japan without appropriate identification. Traces of Bt-10 were found in corn in 2005 that had not been documented as genetically altered.[75]

Japan's national rules furthermore require "prior informed consent" between importers and exporters of transgenic food stocks. Shipments must provide explicit documentation that genetically altered products are included. Scientific experiments need to be certified as safe by research institutes or the Ministry of Education, Culture, Sports, Science, and Technology if the associated risks are high.[76]

A detailed survey analysis of Japanese public opinion found that support for genetically altered food dropped between 1997 and 2003. In addition, by 2003, 50 percent of the Japanese public indicated that "they don't trust their government or biotech companies." This lack of trust helps to explain why Japan has issued tougher rules on genetically modified foods than other Asian countries, and why its officials have been so cautious about commercializing genetically engineered food.[77]

Not surprisingly, given this trend, the Japanese government in 2003 enacted a new "Law Concerning the Conservation and Sustainable Use of Biological Diversity through Regulations on the Use of Living Modified Organisms." This legislation established rules for the import of transgenic food, mandated detailed labeling and documentation for genetically modified food coming into the country, and created regulations for the release of genetically modified plants into the environment.

One major factor in regulatory policies across Asian countries is the variation of public opinion. According to national surveys, consumers in Thailand and India are more favorable toward genetically modified foods than is the case in Japan, Australia, and New Zealand. When asked whether they approved of genetically engineered organisms designed to produce tomatoes with better taste, 83 percent of Thais, 73 percent of Indians, 69 percent of Japanese, 54 percent of Australians, 49 percent of New Zealanders, and 35 percent of Russians were supportive of the new food product.[78]

Conclusion

To summarize, compared to the United States and many Asian nations, European countries have the strictest rules on genetically modified food, although the European Commission is starting to relax its rules in this area. A number of European countries have either outright national bans on genetically altered food or de facto moratoriums on the commercial sale of genetically modified organisms. Nearly all European countries are also likely to have stringent labeling rules, generally for any products containing as little as 0.9 percent of altered ingredients.

In contrast, the United States has the least stringent regulations. There are no mandatory labeling requirements and commercial companies regularly seek and gain government approval to produce and sell transgenic crops. Consumer use of altered foods is quite common, even though people generally are unaware of this fact.

Asian countries are in the middle of this biotech continuum. China and India lean more toward the American enthusiasm for genetically modified food. Meanwhile, Japan is more like many European countries in having mandatory labels and stricter rules on commercializing altered food.[79]

The variation from country to country in regulation of genetically modified food demonstrates the importance of political and economic factors. While the U.S. is likely to employ a "substantial equivalence" test in demonstrating the similarity of engineered crops to their natural counterparts, European countries prefer a higher safety standard. They believe that chemical and genetic similarity do not guarantee the absence of health risks and that scientists need to undertake more systematic studies before genetically modified products are incorporated into the food supply.[80]

Occurrences of damage to food in Europe contribute to the tougher regulation that has developed in some of those locales.[81] Fears over mad cow disease or unlabeled biocrops generate a lot of media attention, much more so than in the United States. This stimulates citizen anxiety and concern, and helps to explain how public opinion leads European countries to take a tougher position on genetically altered foods, compared to the United States and some Asian nations.

The existence of strong nongovernmental citizens' organizations in Europe and the high levels of public opposition to "Frankenstein food" in places such as Germany and France contribute to the policy choices made by these governments. In the United States, the public is not even generally aware that America already sells a large amount of genetically altered food, and that in all likelihood they consume it.

The United States has strong agribusiness corporations that lobby extensively for freedom from stringent regulation. According to analyst Ian Sheldon, the three countries with the weakest rules for overseeing genetically modified foods are the United States, Canada, and Argentina. These nations also happen to represent three of the largest food exporters in the world.[82]

With agricultural trade accounting for a significant portion of exports to the rest of the world, these nations and their respective economic interests are in a strong position to express their point of view and convince legislators that their country needs no labeling requirements and that oversight should be kept to a minimum. Despite citizen concerns about bio-crop safety, American agricultural interests have prevented Congress from holding hearings on genetically modified food and the propriety of engineering basic crops in the United States. This is a testament to their political and economic clout, and the strength of agribusiness in shaping policy discussions.

The presence of contraband trade in prohibited biocrops and the ability of multinationals to "country-shop" for favorable venues demonstrate the challenges of regulating a biotech area where knowledge is mobile and field trials can take place in a variety of locales. If Japan enacts tough rules on genetically modified food, experimentation shifts to China, India, or the Philippines. Or if Germany has serious doubts about altered food, companies can move their operations to Spain or Great Britain.

The mobility of expertise and the difficulty of inspecting every seed shipment make genetically modified food a particularly difficult area to regulate. These represent reasons why the amount of acreage devoted to engineered crops continues to rise dramatically despite concerns in many nations about food and health issues. Companies play nations off against each other, and prevent the most stringent regulations from being implemented.

Chapter 5

Cloning

Biotechnology was turned upside down in 1997 when Scottish scientist Ian Wilmut of the Roslin Institute in Edinburgh announced the successful cloning of a sheep named Dolly. While previously there had been clones of mice, Dolly was the first sheep ever to be copied. Now, clones have been produced of rabbits, calves, pigs, and monkeys, among other organisms.[1]

The proliferation of cloning technology offers great promise for treating medical diseases and furthering international commerce. Because of the possibilities of tissue and organ regeneration through therapeutic cloning, scientists hope that this biotechnology one day will provide transplants or treatments for patients needing assistance.[2] For example, regeneration of skeletal muscles may help those suffering from muscular dystrophy.[3] There also is an economic payoff from cloning. Scientists estimate that animal transplantation has the potential to become a $6 billion global industry when fully developed through xenotransplantation of tissue and organs.[4]

Yet cloning successes have raised a host of ethical issues linked to the propriety of this kind of research and the implications of human cloning. In the pursuit of new treatments for serious diseases, should society allow scientists the freedom to create genetic copies of mammals? And if technology advances to the point where human cloning is possible, should governments restrict or outlaw that practice? Do cloning procedures place researchers in the role of playing God?[5]

Currently, human cloning is banned in Germany, France, and Hong Kong, among other nations. Great Britain allows therapeutic cloning as does China, Malaysia, and Singapore. Most governments have made substantial efforts to regulate research on human cloning and some types of cloning involving other organisms. However, such efforts have not been entirely successful. The mobility of scientists allows them to collaborate with foreign researchers or migrate to locales with fewer restrictions in order to undertake projects forbidden in their own homelands. This helps them bypass the most serious regulations and pursue scientific innovation. Similar to other biotechnologies, it has been a major challenge in an era of globalization and geographic mobility to impose strict rules on experiments undertaken by the science-industrial complex.

Overview of Cloning

The term "clone" was first used in 1903 by H. J. Webber to denote "a colony of organisms derived asexually from a single progenitor."[6] Since then, cloning has evolved into the production of new organisms that are genetically identical to other individuals. Given advances in cell and molecular biology, it has become increasingly possible to clone everything from mice and rabbits to sheep and cows. Some have even speculated that cloning of human tissue is a realistic future possibility.

The most common techniques for cloning involve somatic cell nuclear transfer and embryo splitting. The former involves "removing the nucleus of an unfertilized egg cell, replacing it with material from the nucleus of a somatic cell (i.e. skin or cumulus cell) and stimulating this cell to divide." The transplanted nucleus provides genetic material, while the surrounding egg facilitates the development of an embryo.[7] The latter is "the artificial division of a cell into two or more cells." When the resulting embryo is placed in a female womb, it produces identical offspring.[8]

Reproductive cloning aims to produce an organism that is genetically identical to another individual and placing that organism in a womb for purposes of reproduction.[9] Nonreproductive (or therapeutic) cloning, on the other hand, has more restricted goals. It seeks to "provide compatible tissues and organs for replacement therapy," and is not intended for reproduction per se.[10] Generally, new therapeutic tissues are in existence for a limited time period, until they have stimulated the growth of functioning organs.

Nonreproductive cloning of humans is less controversial and therefore less regulated than reproductive cloning.

A survey of 30 developed nations around the world found that seven countries, including the United States, had no national legislation overseeing any kind of cloning. Of the remaining countries, reproductive cloning is banned nearly everywhere, and 17 nations ban nonreproductive cloning. The latter 17 nations include Australia, Austria, Canada, Denmark, France, Germany, Iceland, Italy, Mexico, the Netherlands, Norway, Peru, Spain, Sweden, and Switzerland. In contrast, countries such as Belgium, China, Finland, Greece, India, Israel, Korea, New Zealand, and the United States have no national legislation on cloning and allow nonreproductive cloning, although some American states have prohibitions within their own jurisdictions.[11]

Reproductive cloning is controversial due to the ethical concerns it raises. Unlike in vitro fertilization, which involves the creation of new life through male and female partners or surrogate donors, cloning bypasses reproduction and allows the creation of new individuals asexually. This short-circuiting of sexual reproduction has spurred critics to question use of the technique on religious and ethical grounds.[12]

The Catholic Church, for example, vehemently opposes both reproductive and nonreproductive copying. It believes reproduction should be based on sexual, not asexual, processes. If humans gain the right to clone themselves, the church fears the technique will be used for genetic engineering, sex selection, and eugenics, none of which is morally acceptable to it. Church officials argue that asexual reproduction runs contrary to religious instruction and violates God's dictum to "be fruitful and multiply."[13]

Debates concerning cloning revolve around arguments over when life actually begins. Many Western religious authorities believe that life begins at conception, and therefore, any human being conceived must be allowed to live, regardless of genetic defects or potential disease. Scientists, however, cite a 14-day threshold as the time required after conception for a backbone and rudimentary human organs to appear in a fetus, at which point the fetus can be considered human. As pointed out by ethicists Steven Best and Douglas Kellner, are "5 day-old embryos in a Petri dish" actually human?[14] Religious fundamentalists would argue yes, while many scientists conclude life is so rudimentary at that point as not to constitute actual human existence.

Others oppose human cloning on grounds of inefficiency and ineffectiveness. According to Dolly creator Wilmut, "Cloning is inefficient in all species. Expect the same outcome in humans as in other species: late abortions, dead children and surviving but abnormal children."[15] One virtue of sexual reproduction is it balances the genetic risks between two different gene pools and thereby reduces the danger of deformities. Animal clones tend to be oversized and suffer from "large offspring syndrome" and what biologists call "in-breeding depression," which is the exposure to deleterious genes through close mating with relatives. They are also more likely to suffer from circulatory and respiratory disorders.[16] For these practical reasons, some scientists oppose human cloning on grounds that it never will live up to its stated hopes for disease cure and economic development.

As a sign of how contentious the issue has become, efforts by the United Nations to promulgate a treaty on cloning have failed due to an inability to forge agreement on the subject. A proposal by France and Germany would have outlawed reproductive cloning, but leave rules on therapeutic cloning for future discussion. Another treaty supported by the United States and Spain proposed a comprehensive ban on all kinds of cloning, but it too was not enacted. In 2003, the United Nations Legal Committee voted 80 to 79 to defer General Assembly consideration of a cloning treaty, with support from the Organization of the Islamic Conference. According to Islamic thinking, it takes at least 40 days after conception for a human soul to enter a fetus. Therefore, it is not immoral within that timeframe to pursue therapeutic cloning for purposes of tissue or organ creation.

However, in 2004, the General Assembly rejected this approach and adopted a one-year delay instead.[17] Later, on an 84 to 34 vote with 37 abstentions, it passed a nonbinding resolution calling on member states to ban all forms of reproductive and therapeutic cloning. This resolution was pushed by the United States and predominantly Catholic countries against the wishes of European and Asian nations such as Great Britain, the Netherlands, and South Korea (which preferred a partial ban).[18] Even though this motion expressed the overall sentiment of the General Assembly, the fact that it was nonbinding meant the United Nations still has not passed a formal ban on human reproductive cloning.[19]

In spite of all the legal and political wrangling, most international bodies have expressed support for a ban on human reproductive cloning.[20] Regardless of the strength of religious forces within a given country, nearly all nations worry that human cloning damages the moral fiber of a nation, sets a dangerous precedent for reproduction, and threatens family institutions by allowing asexual forms of procreation. There is no such consensus, however, on nonreproductive cloning. Because of the possibility of medical breakthroughs, though, some nations are willing to undertake experimentation in hopes that scientists will discover ways to regenerate tissue or find treatments for life-threatening diseases.[21]

The American Experience

Unlike the situation with in vitro fertilization and genetically modified foods, the United States favors a restrictive approach to research on human cloning. While there are no formal limitations on human cloning in the private sector, the national government does not allow federal funding of cloning research. Under President George W. Bush, executive orders have limited the use of federal dollars for this purpose and placed sharp controls on the creation of human embryos. This forces scientists to rely extensively on the private sector for research money.

In 2001, the U.S. House of Representatives enacted the Human Cloning Prohibition Act on a 265 to 162 vote. It outlawed all kinds of human cloning, both reproductive and therapeutic techniques, by any public or private entity. It defined human cloning as "human asexual reproduction, accomplished by introducing nuclear material from one or more human somatic cells into a fertilized or unfertilized oocyte whose nuclear material has been removed or inactivated so as to produce a living organism (at any stage of development) that is genetically virtually identical to an existing or previously existing human organism."[22]

However, the Senate failed to endorse this legislation, undermining efforts by President George W. Bush to ban embryonic cloning. After lobbying by scientific associations, medical groups, and biotech firms opposed to such a limitation, senators were unable to reach an agreement on the propriety of therapeutic cloning for medical purposes. Even though a number of religious organizations favored a complete ban, the lack of consensus among scientific and business groups torpedoed congressional action and left the United States without major national legislation on cloning.[23]

Some states, though, have enacted their own provisions. For example, nine states (Arkansas, California, Iowa, Michigan, New Jersey, North Dakota, Rhode Island, South Dakota, and Virginia) prohibit human cloning. Others such as Missouri forbid public money for research on human cloning. With most other states having no formal regulations in this area, the result is a hodgepodge of government regulation at the state level.[24]

A number of major scientific and medical societies, including the American Medical Association, support human therapeutic, but not reproductive, cloning research. Yet the president's Council on Bioethics has expressed support for a ban on "cloning-to-produce-children" and a four-year moratorium on biomedical cloning research. All the members of the council supported a ban on reproductive cloning, and 10 of the 18 members favored the four-year moratorium.[25]

The absence of federal funding for cloning research has led some American researchers to collaborate with scientists in Asia and Great Britain as a way to get around U.S. limits. Since researchers in these other countries face fewer restrictions and more government funding than in the United States, these collaborations help Americans push the boundaries of medical research. One scientist explained the value of international projects by saying, "We don't have to do everything in every country."[26]

Unlike the situation with in vitro fertilization and genetically modified food, American public opinion does not support reproductive cloning. Most polls have found that around 75 percent of the public oppose this technique. The only cases in which disapproval drops involve "abnormalities in embryos (52 percent disapproved in 1993), cloning of embryos for infertility treatment (63 percent disapproved in 1998), and cloning to produce copies of humans for organs to save others (68 disapproved in 2001)."[27]

In contrast, majorities favor therapeutic cloning for medical purposes if it does not result in human birth. For example,

54 percent of Americans in a 2001 Gallup survey expressed support for cloning to aid medical research, while 59 percent in a 2002 Gallup poll said they favored cloning to produce organs for transplant.[28]

Research is moving ahead in regard to animal cloning in the United States. But currently, the U.S. FDA does not allow the sale of food from cloned animals. The International Dairy Foods Association, which represents dairy producers, opposes lifting this ban on grounds that "consumers are not receptive to milk from cloned cows."[29] Indeed, surveys demonstrate substantial public opposition to this kind of food. A survey undertaken by the International Food Information Council found that 63 percent of Americans "would probably not buy food from cloned animals, even if the FDA determined the products were safe."[30] These sentiments limit the ability of the science-industrial complex to find markets for engineered products in America.

European Experiences

In regard to animal cloning, there are few European Union regulations restricting experimentation and commercialization. Scientists are free to clone sheep, cows, rabbits, and other organisms, and to sell the new creations. Indeed, the cloning of Dolly stimulated a wave of animal copies in a number of different species. Even though some observers have expressed concern that cloned animals may come into the food supply, there are no limits yet on the sale of cloned food at the European level.[31]

Some individual countries, though, have taken action on their own. For example, the Netherlands, Sweden, and Norway have banned animal cloning altogether. But Germany, France, and Italy have not done so, except for some rules governing the use of animals in lab experiments. Italy passed a law banning animal cloning, but it was overturned by its courts. Great Britain allows animal clonings when there is "a clear scientific reason to perform them." British scientists must apply for permission to undertake these kinds of experiments and present a strong justification for the research.[32]

With human cloning, Europe is divided between places that ban therapeutic cloning and nations that allow it with some regulation.

Great Britain lies at the more permissive end of the cloning spectrum. It was the first country in Europe to clone a mammal and the first one formally to endorse therapeutic human cloning for purposes of medical treatment, subject to particular rules and regulations. Following the birth of Dolly in 1997, the government created an advisory panel of experts, and requested guidelines from the Human Genetics Advisory Commission and the Human Fertilisation and Embryology Authority.

In 2001, this led to amendments to the 1990 Human Fertilisation and Embryology Act. These amendments, which passed the House of Commons on a 366 to 174 vote, allowed human therapeutic cloning, but banned human reproductive cloning. To insure that the latter was not attempted, scientists were required to destroy all embryo clones created within 14 days of development, the time British ethicists believe life truly begins. In addition, researchers were not allowed to implant cloned cells into a human womb in order to avoid the possibility of reproduction.[33] Finally, the law "allows isolation of hES (human embryonic stem cells) from spare in-vitro fertilization clinic embryos, and the general creation of embryos for research purposes."[34]

As revealed in parliamentary debates, the major goal of this policy change was to use cloning to improve the understanding and treatment of serious medical diseases. A number of scientific organizations and biotech companies expressed views in favor of cloning for medical purposes. Some religious leaders voiced concern about "wasted embryos" that would be discarded through cloning procedures. However, procloning advocates responded by pointing out that "nature is profligate. We do not mourn for wasted sperm and eggs, alive though they are; nor for the three quarters of fertilized eggs that are lost before implant, half of which are genetically impaired. As the Bishop of Oxford has said, 'If every fertilized egg was indeed a soul . . . then, according to these figures, three quarters of heaven would be populated by souls that lived for less than a week.'"[35]

By 2005, the British climate against human cloning had started to abate. In that year, United Kingdom researchers obtained the first license in Europe to create clones of human embryos. Authorized by the Human Fertilisation and Embryology Authority, leading scientists undertook experiments on diabetes sufferers to determine the degree to which replacement cells producing insulin from cloned embryos would bring relief to these patients. This allowed medical staff to test drugs on cloned tissue as opposed to animals, and thereby speed up experimental trials.[36]

Some British researchers have gone so far as to send human skin cells from live patients to South Korea for cloning. "If we do not collaborate with them it could take us several years. . . . Our patients don't have long. I could not justify not working with the Koreans," a British scientist pointed out.[37] These cloned materials are used to form body tissue and body parts, and shipped back to Great Britain for actual deployment. The hope is that these new creations will regenerate cells for dying patients, without the risk of tissue rejection, and therefore speed the healing process.

At the opposite end of the cloning spectrum from Great Britain are Germany and France. Both nations are much more restrictive about human cloning, and have comprehensive bans on reproductive and therapeutic cloning. In 1991, when cloning technology was in its infancy, Germany banned all forms of cloning research and medical treatment. And since then, it has supported a United Nations treaty calling for a full ban on human cloning.

The country's history of eugenics, which contributed to a public opinion hostile to genetic manipulations, encouraged political leaders to take a tough line against this type of biotechnology. Rather than allow certain kinds of experiments or medical trials, Germans simply banned research in this area altogether and therefore avoided any type of moral uncertainty over the endeavor. Germans were suspicious of cloning, and there was no strong biotech industry pushing for freedom in this area. Furthermore, party leaders did not buy the argument that cloning was the tip of an economic goldmine in trade and commerce.

In 1994, France enacted a ban on "any kind of manipulation or experimentation" on human embryos that since has been applied to cloning research.[38] It extended this ban in 2004 and labeled human reproductive cloning a "crime against the human species" that would be punishable by up to 30 years in prison and a fine of $9.3 million. This tough legislation also made therapeutic cloning illegal for researchers, punishable by up to seven years in jail and a fine of $100,000.[39]

Similar to Germany, France does not have a substantial biotech sector. Its biotech companies are not particularly large and therefore lack the political and economic clout of biotech companies in some other countries. The public holds traditional ideas about genetic modifications to human DNA and is not very sympathetic to claims that this technology is important for health care or economic development. The result is a national political climate that bans cloning human embryos and makes it impossible for scientists to undertake experiments even for purposes of medical research.

In looking at factors that distinguish national responses on human cloning, differences in the clout of the biotech industry, public opinion, and religious and cultural values explain the major variations. Among European nations, for example, the biotech sector is weaker in France and Germany than in Great Britain. While there are many biotech firms on the continent, their market capitalization (and therefore their economic and political clout) is only one-third of that found in Great Britain. The same can be said when comparing these nations to the United States, where the biotech industry is even stronger.

The small size of the biotech industry in many European countries affects the politics of cloning and how different countries respond to therapeutic cloning.[40] In nations where there is a substantial biotech sector, it is harder to ignore arguments about the economic potential of the technology, especially as it is related to medical research. Firms will point to trials they have underway and the possibility that successful experiments will create improvements for those who are suffering from dreaded diseases.

In addition, there are substantial differences in public opinion across European nations about reproductive versus therapeutic cloning.[41] According to a 2003 EOS Gallup Europe survey, a majority of people in European Union nations favor therapeutic cloning, but oppose reproductive cloning. For example, on a question of whether they agree or disagree with reproductive cloning (defined as the identical reproduction of human beings), 5 percent of respondents in European Union countries agreed, while 93 percent disagreed. But on the issue of therapeutic cloning (defined as the identical reproduction of human cells), 55 percent agreed and 43 percent disagreed.[42]

In this survey, views regarding therapeutic cloning, in relation to national policy, were broken down by individual country. Whereas 79 percent of Spaniards and 72 percent of Portuguese agreed with this kind of cloning, the Germans and the French were less likely to support nonreproductive cloning. Only 43 percent of Germans and 57 percent of French agreed that therapeutic cloning was a good thing.[43]

When combined with cultural attitudes about eugenics and genetic experimentation, these citizen sentiments help us to understand why some countries have been more permissive in their cloning rules compared to other countries. The combination of weak biotech firms, unsympathetic public opinion, and cultural values that fear medical testing makes it difficult for nations like Germany and France to undertake research on human therapeutic cloning.

Asian Experiences

Asia exhibits a wide range of views on cloning, from nations that ban all cloning, such as Hong Kong and the Philippines, to those that aggressively pursue the technology for therapeutic purposes, such as Malaysia and Singapore. As pointed out in previous chapters, many countries in this region have religious traditions where there is less concern about cloning, at least for the first few weeks after conception. With many Eastern religions placing less emphasis on the dominion of humans over everything else on the planet, and different ideas about when a fetus becomes a human being, there are fewer domestic constraints on human therapeutic cloning.

In addition, public opinion plays a weaker role in Asian democracies than in their American or European counterparts. In Asian democracies, there is more deference to expert rather than public opinion, especially on matters of science and technology. These topics are seen as technical and complex, and therefore, as areas where mass public opinion is less relevant. The result in many locales has been greater autonomy for scientists and less of a desire for governments to restrict cloning technology.[44]

Among Asian nations, China lies at the more permissive end of the continuum. In 1995, Chinese researchers successfully cloned a goat. This followed the copying of pigs, rats, and rabbits. Animal cloning in China is not controversial and there are few regulations that govern it.[45] And in 2003, China's Ministerial Regulations put forward rules legalizing human cloning for therapeutic treatments based on guidelines from Great Britain. The law is liberal compared to many other nations. It allows therapeutic cloning for medical research and does not prohibit "cross-species fusions" known as chimeras. Researchers at Shanghai Second Medical University have experimented with the implantation of human cells into rabbit eggs that have their genetic nuclei removed. As long as the new embryos are not grown beyond two weeks, Chinese government rules allow this procedure.[46]

Not only are its rules relatively liberal, China's weak enforcement of existing regulations gives researchers additional leeway. According to industry observers, "Chinese institutions are very aggressive in genetic research and regulation is lax."[47] This liberality helps scientists push the envelope of innovation, and gives Chinese researchers considerable freedom in experimentation. Religious or public opinion constraints that make it difficult for researchers in other places to undertake research in this area are not very strong in China. This creates an atmosphere that is quite conducive to biotech innovation.

Korea also allows nonreproductive forms of cloning. The government has a Bioethical Committee that reports directly to the prime minister. This committee promulgates rules for cloning research, funds projects, and sets the parameters for what scientists can and cannot do. Unconstrained by religious groups that limit cloning in other countries or government concerns about bioethics, Korean researchers have employed cloning to match patients with stem cells.[48] The Ministry of Science and Technology provided 2 billion won (or $1.9 million) for this research.[49]

The Indian Council of Medical Research meanwhile has ruled that "research on cloning with intent to produce an identical human being, as of today, is prohibited." However, its policy "opens the door to therapeutic cloning considered on a case-by-case basis by the National Bioethics Committee."[50] Similar to China, its scientists are following United Kingdom guidelines on the use of cloning research. Even though it has a variety of religions within its borders, there is little opposition by various Indian religious authorities to therapeutic cloning.[51]

The fact that Great Britain played a prominent role in the history, both of China and India, illustrates why each country turned to Great Britain for help in formulating cloning guidelines. Rather than look to the European continent or the United States, China and India saw Great Britain as the model of how to proceed on cloning. The fact that Great Britain has relatively liberal rules and that China and India do not face the domestic constraints of countries such as Germany and France help to explain why these Asian nations have been leaders at pushing the frontiers of cloning research.

At the other end of the Asian cloning spectrum is Japan. It has a complete ban on reproductive cloning and a de facto prohibition on therapeutic cloning. To make sure scientists know the government means business, there is a 10-year prison sentence for researchers who violate the cloning ban.[52] According to observers in that country, Japan is very conservative about anything related to medical or genetic experimentation. "Organ transplantation is still rarely done," stated one Tokyo medical professor.[53] In 1968, Japan had its first and only heart transplant. But when the patient died shortly thereafter, the doctors who performed it were accused of murder. The controversial nature of this particular case meant that it took another 30 years until Japan passed legislation authorizing organ transplants from brain-dead patients. This demonstrates the slow pace of biotechnology in that nation.[54]

As an ally of Germany during World War II and an invader accused of atrocities in Korea and China during that war, Japan's approach to cloning follows that of the Germans. Japan is very cautious about genetic experimentation, and its twentieth-century history taught it to be careful of unsupervised medical trials. The result is a national policy that is more restrictive than most other countries in the Far East.

Conclusion

In *Star Wars Episode II: Attack of the Clones,* military planners use clones to serve as warriors for intergalactic battles. This space-age fiction epitomizes for many people the risk of human cloning technology. Media depictions of cloning often promulgate a view of science creating a "brave new world" of human clones. Around the time of Dolly's creation in 1997, *Newsweek* ran a cover story entitled "Can We Clone Humans?" that showed a picture of three identical babies standing in lab beakers.[55] The story asked whether copying humans would lead to a loss of uniqueness and the mass production of human beings, and reflected deep public worries in some places about depersonalization and scientific experimentation. The violation of human reproductive norms through asexual reproduction and the loss of human dignity and identity lead many to oppose all forms of human copying.

Between citizens who are worried about human cloning and media portraits that emphasize cloning as Frankenstein science, it is hardly surprising that many nations have placed sharp restrictions or outright bans on reproductive and therapeutic cloning. Typically, countries that have strong religious organizations opposed to cloning techniques have complete bans on all kinds of cloning, while those that do not (including some nations in Europe and Asia) oppose reproductive, but not therapeutic, cloning.

The combination of religious opposition, unsympathetic public opinion, and a weak biotech industry in many nations makes for an atmosphere that is not conducive to cloning technology. Church officials rail against human cloning and express concern about the loss of human dignity and individual identity. This sometimes is enough to make government officials pass laws outlawing cloning.

In other places, where religion is not strong or not opposed to therapeutic cloning, or where there are strong scientific or biotech companies that lobby government, legislation tends to be weaker with more allowances for cloning to advance medical research. This gives the public sector more freedom to fund research without worries about a backlash from moral elements within the society.

Even in nations that have placed sharp limits on therapeutic cloning, there is no guarantee in an era of globalization that these laws actually will restrict cloning. As pointed out by writer Elisabeth Rosen, there is a dichotomy between dollars and Dolly. "Banning animal cloning in Europe would not necessarily mean freeing its territory from its expected consequences," she said. "It could result in industrial migration overseas, loss of employment and profits that would be geopolitically redistributed."[56] Rather than acting as a constraint on biotech innovation, the mobility of knowledge across national borders in the contemporary period makes it difficult to stop medical experimentation in this area.

Although it is uncertain how effective this experimentation will be, the global nature of cloning research for medical purposes and the difficulties facing countries that want to limit certain practices within their own borders makes the area challenging to regulate. It no longer is possible to keep knowledge within a single country and prevent scientists from collaborating across national boundaries. The globalization of scientific innovation and the rise of a global science-industrial complex make it nearly impossible to control biotechnology within individual nations. If one country seeks draconian restrictions, its scientists will develop partnerships with peers in other lands, and use those collaborations to engage in research that is prohibited at home. This poses a continuing quandary for nations with ethical qualms about human cloning.

Chapter 6

Stem Cell Research

Regardless of the nature of the national political climate, most countries find stem cell research extremely controversial. Because experiments on human embryos are seen as threatening life and human dignity, and as a number of religious organizations have taken strong stances against research in this area, many governments have placed sharp restrictions or outright bans on public funding of human embryonic stem cells. The well-organized character of the opposition and the fundamental nature of the ethical considerations have turned this policy area into one of high visibility and high conflict. Generally, this combination of qualities is the ideal recipe for a restrictive biotech policy approach.

At the same time, with the global market for stem cell research and tissue transplantation expected to reach $10 billion by 2013, some nations are investing major financial resources in this area. This is particularly true in the case of Asian countries such as China, South Korea, and Singapore. They are less constrained by domestic political and religious opposition, and see stem cell research as a policy domain where they can leapfrog Western nations slowed by intense social and political pressures.[1]

The combination of religion, politics, and economics has turned this biotechnology into one that has different dynamics from other areas. Unlike in vitro fertilization or genetically modified food, which remain the domain of scientists, experts, and industry groups, conflict over stem cell research has expanded into the world of politicians, voters, and news media. This expansion in the scope of conflict has increased the amount of public discussion and generated considerable disagreement over the proper course of action.

As noted by Matthew Nisbet in his study of media coverage of biotech policy from 1975 to 2002, "Policy debate has also moved from administrative and technical arenas to overly political arenas such as the White House and Congress." In Nisbet's eyes, this reflects the moral character of this issue and rhetorical frames based on "science as moral panic and political game."[2]

The end result has been more systematic regulation compared to other areas of biotech research. Some nations ban work in this area, while others prohibit government funding of stem cell research. Although most nations have been cautious in pursuing research on embryonic stem cells, there is variation around the world in how this biotechnology is handled by the public sector. Nations that are most supportive of stem cell research include Great Britain, China, Singapore, and South Korea. In contrast, Germany, Austria, and the United States are more restrictive in their policies and much less willing to use public funding to support research in this area. By comparing national policies on stem cell research, one can gain a greater appreciation of the factors that impede this kind of technology.

Overview of Stem Cell Research

Using human cells to generate tissue is not a new undertaking. For several decades, scientists have been transplanting adult cells from bone marrow into patients for the treatment of leukemia, cancer, or other diseases. The advantage of this procedure is that adult cells generate new bone or blood to replace damaged or diseased cells. This helps the sick battle both cancerous cells and the side effects of chemotherapy.[3]

However, such traditional approaches suffer from two disadvantages. First, there is a need to match donors and recipients. Those receiving cell transplants require a similar makeup to avoid rejection by the new host. In addition, adult cells are limited to a specific function. Blood cells can be transfused to supplement a person's blood supply and cartilage can be transplanted into a new patient. But doctors cannot turn blood cells into cartilage, or cartilage into bone. This lack of versatility limits the treatment options available to needy patients.

In 1998, scientist James Thomson at the University of Wisconsin was able to identify human embryonic stem cells that could grow into many different cell lines. Embryonic stem cells are the "biological building blocks that serve as the common ancestry of all 210 different kinds of tissue in the human body."[4] These cell lines can be removed from fetuses at an early phase of development, from human placenta, and from blood collected from umbilical cords. After identifying and isolating these cells, they can be cultured into skin, blood, heart, or other kinds of human tissue, and used for medical treatments.[5]

Timing is crucial to this procedure. When a female egg is fertilized by a sperm, the resultant zygote is a single cell that carries the genetic code for the entire body. Almost immediately, the zygote starts dividing into different cells. About nine days after gestation, around 200 cells form a sphere called a "blastocyst." It has an outer layer that evolves into a placenta and an inner layer of "pluripotent cells" that form the basis of stem cells.[6] These pluripotent cells must be removed quickly before chemical signals are received that turn them into specialized tissues, such as blood, heart, or skin cells. If this is done between 9 and 11 days following conception, scientists can use the generic stem cell lines obtained to generate new tissues and organs.[7]

The controversy surrounding this process involves the destruction of embryos and use of cells from terminated pregnancies to generate stem cells. Similar to the situation with cloning, in some procedures, removing stem cells destroys the embryo. Christian authorities, who feel that human life begins at conception, object to embryonic stem cells that originate from procedures they regard as immoral and unethical.

In contrast, many in the Moslem and Judaic faiths believe "ensoulment" does not take place until 40 days following fertilization. Rather than thinking life begins at conception and therefore cannot be altered or manipulated at any point after an embryo is conceived, they view actual life as beginning at a later point in fetal development. Followers of Islam and Judaism are therefore not as worried as fundamentalist Christians are about the morality of this procedure in the days before the soul enters the body.[8]

Other ethicists outside the religious area fear the specter of scientists growing hearts, lungs, kidneys, or bones from human embryonic stem cells. If laboratories turn into tissue factories, ethicists ask whether humans are getting too close to playing God?[9] Leon Kass, chairperson of the President's Council on Bioethics, refers to this as "the wisdom of repugnance" and claims new biotech breakthroughs are "repulsive," "revolting," and "grotesque."[10] These kinds of moral dilemmas have led to intense political lobbying over cloning and stem cell research.

Advocates of stem cell research respond by pointing to the impossibility of knowing when life begins and the potential advances in medical treatment. Douglas Melton, a stem cell researcher at Harvard University, claims ethics arguments are "all about differing religious beliefs. I don't believe I have the right to tell others when life begins. Science doesn't have the answer to that question; it's metaphysical."[11]

In addition, scientists believe stem cells represent a source of treatment for diseases that result from damaged cells or ruined tissues. For example, Parkinson's disease involves nerve cells that fire indiscriminately, leaving the person with uncontrollable hand or leg tremors. Alzheimer's results from the breakdown of brain neurons. These and other maladies such as spinal cord injuries, diabetes, and multiple sclerosis may be treatable if new cells can be generated to replace lost or damaged tissues.

The possibility of these kinds of medical breakthroughs encourages scientists to move ahead despite ethical concerns. If new treatments can repair damaged nerve cells or body organs, they are seen as a reward worth the risk. Helping people overcome diseased tissues and organs is ground for optimism. As long as experimentation is subject to systematic review, supporters of this kind of research argue that they should be allowed to push the boundary of biotechnological innovation.

The American Experience

The U.S. government has been quite cautious in how it handles stem cell research. Unlike the cases of in vitro fertilization and genetically modified foods, but somewhat similar to how it has dealt with cloning, the United States favors a restrictive approach to this kind of biotechnology. The national government currently has sharp limits on public funding of human embryonic stem cells and is not eager for scientists to push research very far in this policy area.

During the Clinton presidency, national policy prohibited the creation of embryos purely for the purpose of removing stem cells. Embryos that were leftover from in vitro fertilization clinics could be used for medical experimentation as long as the patient voluntarily signed a waiver form agreeing to that purpose. There could be no money or other kinds of incentives given to encourage people to donate embryos for this purpose. There was some limited federal funding for research, but it was subject to strict regulations.[12]

However, on August 9, 2001, in a nationally televised address, President George W. Bush fundamentally altered public policy. He banned federal funding for stem cell research, and announced that his administration would not allow scientists to create any new embryonic stem cell lines beyond the 78 that existed prior to that date, about a dozen of which were usable.[13] As President Bush stated during the speech, "This issue forces us to confront fundamental questions about the beginnings of life and the ends of science. It lies at a difficult moral intersection, juxtaposing the need to protect life in all its phases with the prospect of saving and improving life in all its stages."[14]

The result was that American researchers had to secure funding from private sources and build separate facilities not supported by federal money. According to professors at Harvard University, "Because of administration policy, we had to set up this whole new laboratory that was separate from everything else here at Harvard. And we had to separate the money in a really scrupulous way. We have an accountant who makes sure that not a penny of federal funds goes to embryonic stem cell research. We have separate everything—light bulbs, computers, centrifuges."[15]

Some members of his own party disagreed with Bush's stance. When in 2005, the proposed Stem Cell Research Enhancement Act, which would resume federal funding of stem cell research on unused in vitro embryos, was voted on in the U.S. House of Representatives, 50 Republicans broke with Bush to support the act on a 238 to 194 vote.[16] However, the status quo was maintained when the Senate failed to act on the legislation.[17]

The effect of the Bush administration's ban was the shift of stem cell research to off-campus, commercial sponsorship from private industry. For example, experimentation at the University of California moved to an off-campus facility financed by the private firm Geron. Because the research was off-campus and under private financing, there were fewer regulations than when the federal government funded such research. One scientist noted that "the Bush policies mean that, in the U.S., so long as you have private money, you can do everything—buy embryos, create embryos, do it in your basement."[18]

Ironically, by restricting federal funding, the anti-stem cell policy expanded the power of the science-industrial complex. University professors were forced to turn to the private sector for funding of their work. Many of these projects were outside the public eye and not subject to federal control. For research that held commercial potential, the option for private funding gave scientists greater leeway in undertaking the research than would have been the case under a tightly regulated government jurisdiction.

Recognizing this reality, the National Academy of Science in 2005 recommended new guidelines for stem cell research. According to the proposal, universities should create review boards to oversee research on human embryonic stem cells. In addition, the guidelines stipulate that there should be no research on embryos after 14 days, the time scientists believe rudimental skeletons and organs are formed in human beings.[19] Put plainly, the Academy argued that research should move ahead, but be subject to greater self-regulation.

As a result of the federal ban on funding stem cell research, some states have moved into the breach with substantial funding projects.[20] California, for example, has agreed to spend $3 billion over 10 years on stem cell research. Following a 2004 referendum in which 59 percent of California voters approved a ballot measure to use human embryonic stem cells to fight illnesses, the state established an Institute for Regenerative Medicine to distribute the money.[21] Research progress has been slow, though, because legal and partisan wrangling has made it difficult to set up the institutional process for resource distribution.

When Democrats regained control of the U.S. House and Senate in 2006, their first order of business was to pass legislation authorizing federal spending on stem cell research. After years of inaction by the Bush Administration, Democrats saw the potential medical payoffs of stem cell research as a popular issue for their party. They pushed stem cell research as a way to reap future health care benefits.

With stem cells attracting considerable news coverage, citizens are paying closer attention to stem cell research. In 2001, national surveys found that only 20 percent of Americans said they were following the issue closely. This indicates the relative paucity of attention paid to this biotechnology. After Bush's national speech on the subject, however, 60 percent indicated they were paying attention.[22]

The dramatic rise in public interest in this subject reflects the outpouring of media coverage of the issue, the higher visibility of the area, and the partisan differences that exist between stem cell proponents and opponents. With extensive conflict centering on moral issues, the public is devoting greater attention to this subject and forming impressions of what they think should happen. When asked how they felt about stem cell research, 63 percent said they favored stem cell research and 33 percent opposed it when the process relied on extra embryos produced by fertility clinics. Support drops, though, when the question moves to research embryos (46 percent) and cloned embryos (28 percent).[23] A majority of Americans (54 percent) say they oppose the fertilization of human eggs specifically for the purpose of generating stem cells for research.[24]

Americans who attend church regularly are the least likely to support the use of embryonic stem cells for medical research. Only 38 percent of weekly churchgoers found stem cell research morally acceptable, compared to 56 percent who attended "nearly weekly" and 67 percent who never participated in church services.[25] These figures show the close intersection of politics and religion in the United States when it comes to public attitudes regarding stem cell research.

European Experiences

Within Europe, Great Britain has the most liberal policies in regard to stem cell research.[26] In 2000, it authorized the creation of human embryos via in vitro fertilization or the transfer of cell nuclei for research purposes through the Human Fertilisation and Embryology Authority. Licenses for this research must be obtained through the authority.[27] The approval of this new policy came following a Department of Health report that endorsed an expansion of research on human embryos. The report advocated projects that would extract embryonic stem cells from embryos five- to six-days-old, either from in vitro fertilization or nuclear transfer techniques.[28]

In 2004, the British government became the first nation to allow scientists to create human embryos for the purpose of harvesting stem cells. All lines generated through these procedures would be deposited in the UK Stem Cell Bank, a $4.7 million storage facility financed by the national government.[29] A 2005 Stem Cell Initiative promised 2.5 billion pounds (or $4.4 billion) to support stem cell research over the following decade. Consequently, newspapers reported that "restrictions on research in the US and elsewhere mean academics from all over the world are coming to work in Britain."[30]

Britain's liberal government policy reflects general public support for stem cell research. Opinion polls in Great Britain find that 57 percent of Brits feel stem cell research is morally acceptable. By a two-to-one margin, people are more likely to find this kind of research acceptable than not. Stem cell support is strong across a wide variety of backgrounds and beliefs.

In addition, Great Britain has the strongest biotech industry in Europe. Despite its relatively small size, Great Britain has 48 publicly listed companies that are responsible for "60 percent of the sector's value [in Europe] and for three-quarters of the products in late-stage trials."[31] These companies have actively lobbied the central government for policies that facilitate biotech research. They argue that stem cell research has great medical and economic potential for helping people.

Other European countries also have a favorable public opinion climate for biotechnology. In 2004, for example, Swiss voters approved the authorization of a new law that legalized stem cell research on human embryos by a 66 to 34 percent margin. The referendum came to the fore because of a controversial law passed the previous year by Swiss parliament that would have allowed stem cell research, but was opposed strongly by religious organizations. One of the 2003 bill's opponents complained that "the law on stem cells withdraws the right to life from living, defenseless humans. This kind of idea always leads to the destruction of humanity."

To protest that legislation, critics collected signatures to place the measure on the ballot. But following active campaigning, voters sided with those favoring stem cell research. According to news accounts, economic interests were important in the referendum. Several biotech and pharmaceutical corporations are located in Switzerland, and argued vigorously on behalf of the new legislation.

At the more restrictive end of the policy scale from the examples of Great Britain and Switzerland are Italy, Ireland, Poland, Austria, and Germany. These countries prohibit research on all kinds of human embryos. They also do not allow research on previously created embryonic stem cells. Unlike Great Britain and Switzerland, these nations have serious reservations about the propriety of this kind of work.

Germany is an example of a nation with more restrictive policies. Through its Stem Cell Act of 2002, this country does not allow research on human embryonic stem cells. The only exception is a loophole that allows stem cell research if the embryos have been imported from abroad before January 2002.[32] This is intended to facilitate research by those who started projects several years ago. As a sign of the seriousness with which these legislative provisions are taken, those who violate the prohibitions of this legislation face up to six years in federal prison.[33]

This legislation was strongly advocated by the Green party when it was in the government coalition with the Social Democrats. Even though scientists and business leaders opposed the strict rules forbidding stem cell research, the then chancellor Gerhard Schroeder supported the law banning the use of embryos for stem cell research in order to keep his political coalition together.[34]

Stem cell research remains controversial among the German general public. After Chancellor Schroeder suggested a relaxation in Germany's anti-stem-cell rules, opinion polls found that 40 percent of Germans favored liberalizing the law in order to find new treatments for diseases, 30 percent opposed liberalization, and the rest were uncertain. This is much lower support for stem cell work than the levels of support found in Great Britain. Not helping Schroeder's stance was the fact that his comments were criticized by the Green party and the Christian Democratic Party as well as some members of his own Social Democratic political party.[35]

Initially, France and Spain had restrictive policies in regard to stem cell research. Similar to the areas of cloning and genetically modified food, these nations were wary of biotech innovation and banned medical research involving human embryos. Dating back to France's 1994 bioethics law, the country had a strict prohibition on embryo research. The opening section of one law stipulated that "the artificial creation of embryos in vitro for research purposes is prohibited."[36]However, in 2004, both France and Spain enacted legislation allowing research in this area, subject to government review.[37] This moved each nation closer to the more liberal part of the European spectrum. Researchers could undertake some types of stem cell work as long as it was reviewed and approved by government agencies. The hope was that advances in developmental biology would provide help in treating major diseases and would serve as an engine of economic development in each country.[38]

Internal divisions within Europe have complicated the ability of the European Commission to produce a cohesive policy for stem cell research. In 2003, the European Parliament voted 298 to 241 with 21 abstentions to adopt a policy allowing the use of and funding of research on human embryos to produce stem cells as long as the research took place in the first 14 days after conception and that embryos had been created through infertility treatments. This closely contested vote illustrates the lack of consensus on stem cell research and divisions that exist along religious lines in many countries. Indeed, outside observers concluded that "religion was the key factor in this debate" and claimed the battle was reminiscent of earlier conflicts between Catholics and Protestants at the time of famed religious reformer Martin Luther.[39]

Asian Experiences

As has been true in regard to several areas of biotech policy, one of the advantages Asian nations have over Western ones is that their religious traditions do not inhibit work on stem cells. Neither Buddhism nor Confucianism, for example, opposes research on human embryos in the early days of development. According to experts, "Across Asia, there is little of the conflict with prevailing religious and ethical beliefs that has Western countries hesitating."[40]

In addition, public opinion in the region is sympathetic toward stem cell research. Darryl Macer says that "there is a positive view toward human stem cell research in its potential to save life. Surveys have also found positive views towards science across Asia."[41]

Between the absence of religious opposition and the generally positive public opinion climate, government officials are able to push the boundary of biotech innovation beyond what is possible in many Western nations.

The result of this favorable atmosphere is that many Asian countries (especially China, Singapore, Taiwan, India, and South Korea) have liberal policies on stem cell research. These countries allow the creation of human embryos for purposes of medical research and the transfer of nuclear materials into human eggs. Each authorizes the use of embryos from leftover infertility treatments. Both Singapore and South Korea have national bioethics committees that hold hearings and suggest guidelines for ethical research.[42]

South Korea liberalized its stem cell legislation in 2003. Its National Assembly enacted a Bioethics and Biosafety bill that allows research on embryos that have been kept frozen for at least five years. It also legalizes the use of nuclear transfer procedures for purposes of medical research.[43] The national government placed a big bet on stem cell research by providing scientist Woo Suk Hwang of Seoul National University with $26 million in funding.[44] In 2002, it established a Korean Stem Cell Research Center with an annual budget of $7.5 million. More recently, it announced plans for a new Bio-MAX Institute funded by $50 million in government money with an emphasis on stem cell research.[45]

Initially, this investment appeared to yield major results when Hwang announced in 2005 that he and his collaborators had produced 11 embryonic stem cell lines through successful cloning of a human blastocyst, a ball of 200 cells produced nine days after conception and essentially the human "factory" for embryonic stem cells. While this experiment had been undertaken successfully in animals, the Korean breakthrough appeared to represent the first time scientists had done so in humans. This discovery, published in *Science* magazine, generated front-page news in leading publications around the world.[46] Hoping to gain commercial advantage, Hwang proposed the creation of an institute, The World Stem Cell Foundation, which would "perform the cloning service for any scientist who wishes to establish a research culture of cells from patients suffering from particular diseases." Satellite clinics would be created in Great Britain and the United States. Hwang's technicians would perform the cloning in these clinics and send the cells back to Seoul to be "checked, banked and distributed."[47]

However, the following year, these results were retracted when deficiencies were found in the data. According to an investigation at his university it was found that Hwang and his research team "had split cells from one patient into two test tubes for the analysis—rather than actually match cloned cells to a patient's original cells." This meant that nearly all of the supposed new lines actually came from the original patient. It was a costly and embarrassing setback for the Korean stem cell effort.[48]

Embryonic stem cell research is the cornerstone of Singapore's $2 billion National Biomedical Science Strategy that the government initiated in 2000. Its Agency for Science, Technology, and Research is spending around $7.3 million on grants to university scientists, and the government runs a new Biopolis life sciences center. A leading researcher, Ariff Bongso, has developed a number of new human embryonic stem cell lines generated without "mouse feeder layers" and given these materials to a company ES Cell for commercialization. Currently, that enterprise has sold over 140 stem cell lines to scientists around the world, second only to the Wisconsin Alumni Research Foundation.[49]

Singapore is an excellent illustration of how "country-shopping" and "scientist-buying" operate in the stem cell field. Its ES Cell has recruited leading British and American scientists to work on the commercialization of stem cells. In some cases, these scientists have managed to undertake projects in Singapore that would have been impossible in their home countries, due to either legal or financial limitations.

In addition, the U.S. Juvenile Diabetes Research Foundation has provided major grants in Singapore to fund Spanish and American researchers unable to pursue stem cell experiments in their native lands. According to its chief scientific officer, the foundation is supporting stem cell research in Singapore "because there is excellent science, a good environment, and really strong support for work that can't be done in the U.S."[50] This demonstrates how the globalization of biotech innovation has liberated scientists from the limits of national boundaries and freed them to push the lines of innovation.

China is spending around $24 million a year on stem cell research.[51] Overall, the Chinese government plans to boost science and technology research funding from 0.48 percent of its GDP in 1995 to 1.5 percent in 2005 and 2.5 percent in following years.[52] The government issued embryo research regulations in 2003 that are quite liberal. The only restriction on cloning embryos for research purposes is that "scientists comply with the principle of informed consent and informed choice when obtaining embryos from IVF clinics or fetal tissue from aborted fetuses."[53] This is much less restrictive than the case in many Western nations.

One American-trained Chinese researcher described her transition to research in China saying simply, "Here I can work on this important problem, and there [in the United States] I couldn't." Chinese scholars have even managed to undertake research projects using stem cells taken from bone marrow of aborted fetuses, some from second trimester terminations. This is a research approach that is banned by many American states and forbidden from receiving federal money in the United States.[54]

Chinese scientists are setting up companies to commercialize stem cell research through clinical experiments on human beings. One firm spun off by researchers at Beijing University is experimenting with stem cell therapies for damaged human corneas. According to its business plan, "This has been tried in three patients with encouraging results. Their expectation is two to three years for pre-clinical work and a further two years for taking these cells for use in the clinic."[55]

Another company raised $30 million from Western investors to investigate preservation of umbilical cord blood for stem cell research. The business used this money to set up a cord blood bank in Tianjin, China. This allowed patients to store blood with the bank and receive future stem cell treatment.[56]

In general, there are fewer qualms in Asian countries about stem cell experiments on human beings. This is universally prohibited in Western nations, even those with more liberal stem cell policies. But in China, scientists use stem cells in clinical therapies and watch to see what happens with patients.[57] This trial-and-error approach is weakly regulated by the government, even though researchers are aware of the possible dangers in such medical treatments.[58]

Clinical trials were stopped in the United States after a patient died in 1999. Because China's regulations are weaker than in America, Chinese scientists are providing gene therapies not available elsewhere. For example, the Beijing-Haidian Hospital offers treatments of Gendicine, a gene therapy drug thought to destroy tumor cells. Patients from 22 different countries have come to that hospital to receive cancer therapy.[59]

India also has an aggressive stance regarding use of human embryonic stem cells in clinical medical trials. Despite formal prohibitions nearly everywhere else, Indian researchers have been conducting such trials for several years. According to government officials, "Today, anyone can offer stem cell treatment as no permissions have to be sought."[60]

For example, the privately operated L. V. Prasad Eye Institute in Hyderabad has treated 240 patients with cornea problems through transplanted stem cells. Other private hospitals in India have used stem cells to treat the heart, liver, and pancreas. Some doctors even have used stem cells for treatment of strokes, cerebral palsy, and muscular dystrophy. Although the country is moving toward some regulation on these practices, government officials admit that current monitoring attempts are woefully inadequate.[61]

India's Department of Biotechnology expects to spend about 30 percent of its 5 billion-rupee (or $111 million) budget on stem cell research. This amounts to around $33 million. The money would be spread out among various university, government, and business research labs.[62]

Similar to its stance on other biotechnologies, Japan is more restrictive than other Asian nations in its approach to human embryonic stem cells. It permits research only with embryos that have been leftover as a result of in vitro fertilization. Following national legislation enacted in 2001, Japan allows stem cell research with excess embryos derived from married couples undergoing infertility treatments, with written consent and no compensation.

Japan's Ministry of Education, Culture, Sports, Science and Technology has ten pages of guidelines designed to control the use of stem cells. Interested scientists must submit to reviews both in their home institutions as well as from the national government. Approval is not automatic, as use of each cell line in the research plan is discussed and assessed by authorities. Almost all of the stem cell research in Japan is funded by the national government.[63]

Since 2001, outside advisory panels in Japan have recommended increased liberalization, such as allowing the creation of embryos for purposes of medical research. Scientists there have complained that they are falling behind other Asian countries and that the reviews are too detailed, time-consuming, and intrusive. But unlike other nations in the region, the Japanese government has been very cautious about moving ahead in this biotech area. It worries about the ethics of stem cell research and the propriety of genetic engineering work being undertaken by scientists and commercial companies in Japan.[64]

Conclusion

In reviewing national stem cell policies, it is clear there is considerable variation in how different countries handle this issue. Although most nations are careful to oversee stem cell research, China, South Korea, Singapore, and Great Britain have the most permissive policies, while Austria, Germany, France, and the United States are more restrictive about funding this kind of work.

The restrictive policies in the United States and some European nations has led other countries to raid the West for talent willing to work under less stringent conditions. "We have spent the last 50 years in this country building a biomedical research enterprise that's the envy of the world, but with stem cell research, we are giving that lead away," complained one American scientist. Due to funding and regulatory limitations, some American scientists are going to South Korea, China, and Great Britain where there are fewer limits on what they can study. "Science is moveable. Centers of regenerative medicine will spring up outside of this country," warned an American biologist concerned about the funding limitations.[65]

Even Japan, one of the most cautious Asian nations, is raiding the United States for research talent. A new government-funded Center for Developmental Biology in Japan recently lured two prominent American scientists from a private biotech firm in Massachusetts to the new facility in Kobe. With a $45 million budget, the Center provides facilities and funding that are unavailable in the United States.[66]

This type of "country-shopping" by scientists helps them avoid government efforts to restrict biotech innovation. Because knowledge is mobile and not constrained by geographic boundaries, members of the science-industrial complex migrate to locales with fewer restrictions in place. As pointed out by writer David Resnik, "Prohibitions on the use of government funds can simply force controversial research into the private sphere, and unilateral or multilateral research bans can simply encourage multi-national companies to conduct research in countries that lack restrictive laws."[67]

Furthermore, there is a problem with privately funded research. Experimental projects that are undertaken through commercial sponsors generally have fewer institutional oversight mechanisms than those that are publicly funded and there is less sharing of the research data and results. Government grants subject researchers to conflict-of-interest rules as well as institutional review boards that are stricter than those found in industry. Public sector rules have "more stringent safety, informed consent, record keeping, and data monitoring standards" than those in industry.[68]

For these reasons, there are risks with "privatized biomedical research." To the extent that a lack of public funding and strict government rules drive biotech innovation out of the public sphere into the private sector, there is less oversight of and control over biotech research, and greater power for the science-industrial complex. This weakens the ability of governments to control biotech innovation.

Chapter 7

Chimeras

Many cross-species fusions are not controversial from a moral or ethical standpoint. They don't involve matters of life or death. The research does not place humans in the role of playing God. No one protests the mergers, and the creations that are being generated receive little press attention.

The classic case of crossbreeding involves the union of male donkeys and female horses to produce mules. Considered the most successful cross in the history of humankind, this combination created a new entity for trade and commerce and became an engine of economic development in the Old West. The goal of many settlers of western areas in the United States was "40 acres and a mule." Armed with that objective, they crossed the Great Plains and developed new parts of the country.[1]

More recently, cross-species transplantation has allowed pig's valves to be transplanted into human hearts. Scientists have also placed human cells in the liver and pancreas of sheep in order to make these animal organs more suitable for cross-species transplantation. Furthermore, chickens have been raised on eggs injected with special human proteins in order to give them disease-fighting properties.[2]

But when new creations involve genetic crossbreeds based on the living cells of human beings and other animals, the ethical terrain shifts markedly. These animal-human hybrids known as "chimeras" are controversial because they redraw the lines between human beings and animals.[3] Rather than being seen as a technological achievement with few moral ramifications, chimeras evoke strong reactions from a range of different quarters and generate calls for a ban on this kind of research.

As an example, the injection of millions of human brain cells into monkeys raises questions about whether primates are being imbued with human consciousness. As bioethicist Francoise Baylis points out, "We have to be sure we are not creating beings with consciousness." Since there are very different rules, expectations, and religious ramifications for the treatment of humans and animals, Baylis said, creating "ambiguous creatures could lead to 'inexorable moral confusion.'" This is an outcome she feels should be avoided in order to preserve the special status of human beings in the universe.[4]

In this chapter, I look at experiments involving chimeras to determine how they are being handled by various countries. What kind of work is being undertaken, and who is funding these research activities? Although many nations worry about the ethical overtones associated with these kinds of projects, few have rules regulating these endeavors. Indeed, a number of scientists and private companies are moving ahead with work in this area, in hopes of producing medical breakthroughs on major diseases. Chimeras represent a significant indication of the power of the science-industrial complex. Despite ethical concerns over the propriety of this biotech domain, many nations allow this research to take place because of its potential for commerce and medical advances.

Overview of Chimeras

The term "chimera" dates back to the mythical Greek creature Chimera, which was said to have the head of a lion, the body of a goat, and the tail of a snake or dragon.[5] It breathed fire and provided evidence of impending storms or natural disasters. Medieval Christians believed chimeras were an indication of Satan's evil power. Artists sometimes depicted chimeras at the gates of hell taunting lost souls. Indeed, Flemish painter Hieronymus Bosch was famous for incorporating chimeric monsters in his nightmarish fifteenth-century oil portraits.[6]

In modern times, chimeras are produced by scientists in the laboratory by mixing cells from different species. In 1984, a "geep" was produced by combining goat and sheep embryos. There also have been combinations of quail brains and chicken embryos that produced baby chicks that "made sounds like baby quails."[7]

Yet the most controversial combinations are not constructed between animals, but rather human-animal chimeras. The first human chimera is believed to have been produced in 2003 when Chinese researcher Hui Zhen Sheng of the Shanghai Second Medical University fused genetic material from humans and rabbits. For a few days, she allowed the combination of rabbit mitochondria and human skin cells to coexist together before destroying the new creation in order to harvest its embryonic stem cells. Nearly 100 of the 400 embryos survived to the blastocyst stage.[8]

Two years later, a Virginia Commonwealth University scientist, Christopher Kepley, generated international headlines when he and his colleagues at UCLA combined human and cat proteins to treat human allergies to felines. After fusing the cat protein responsible for allergic reactions with a human protein that suppresses those allergens, he discovered that this new combination stopped cat allergies in laboratory mice. Cells with this chimeric protein produced 90 percent less histamine than those without it.[9]

Stanford researcher Irving Weissman, a biologist and cofounder of the commercial biotech firm StemCells Inc., has raised mice with brains that have traces of human brain cells. These cells total one percent of the mice's overall neurons. In this work, he added human neurons to mice brains and found they survived the transplant. But his research team could not determine if the neurons functioned as actual brain material. In order to measure this, he is seeking approval for future experiments that will create mice having 100 percent human brains.[10]

In a related vein, San Diego scientist Fred Gage grew mice with brain cells that originated from human embryonic stem cells. At the 14-day threshold, mice fetuses were injected with 100,000 human cells each. Mouse brains typically have 75 to 90 million cells. A few hundred of the human cells survived as each mouse grew to adulthood and morphed into mouse brains. According to Gage, the rodents' brain reprogrammed the implanted human cells into mouse cells.[11]

The fundamental nature of these experiments across species raises several ethical concerns. First is the moral question of what percentage of an organism's genes has to be human for that entity to be considered a person, and treated as such. University of Chicago biologist Janet Rowley explained that the greatest danger of this work was unknowingly creating a human being in the form of a mouse. "All of us are aware of the concern that we're going to have a human brain in a mouse with a person saying, 'Let me out.'"[12]

Others feel human-animal chimeras diminish human dignity. For example, Cynthia Cohen, a research fellow at Georgetown University's Kennedy Institute of Ethics, believes chimeras "would deny that there is something distinctive and valuable about human beings that ought to be honored and protected."[13] On that basis alone, she concludes there should be sharp limits and extensive oversight on chimera research and implanting human cells, especially brain neurons, into living animals.

Still others worry about the unknown consequences of chimera research. William Cheshire, a neurologist with the Christian Medical and Dental Associations says, "We must be cautious not to violate the integrity of humanity or of animal life over which we have a stewardship responsibility. Research projects that create human-animal chimeras risk disturbing fragile ecosystems, endanger health, and affront species integrity."[14]

Despite these concerns, scientists and business leaders argue that such work helps people with dreaded diseases and opens lucrative markets for medical treatment. "There's a bizarre misunderstanding about the nature of life," says Stanford Law Professor Henry Greeley, an ethicist who oversees Weissman's mouse experiments. "There aren't 'human genes' and 'goat genes.' There are just genes that have different functions."[15] Or to put it differently, the often-cited line between humans and animals is not as clear-cut and distinctive as typically considered. Weissman himself responds to critics by noting the potential for new treatments of brain disorders such as Alzheimer's and Lou Gehrig's disease. He asks rhetorically, "Which of these diseases should we not pursue?"[16]

Because of the health and medical potential of chimeras, many countries have not yet passed laws governing research in this area. Rather than attempting to resolve the ethical concerns and define clearly what is permissible and what is not, policymakers are waiting to see how far advances go before regulations are imposed. But as discussed later, this trial and error approach to ethics places too much trust in the science-industrial complex. The autonomy granted to researchers in this area raises a number of problems for societies around the world.

The American Experience

Unlike some kinds of biotech innovations, chimera research is not restricted by law in the United States. Neither the national government nor state agencies have enacted statutes that control how scientists undertake chimera research or what kind of work they are allowed to complete. The American government does not fund human chimera research, except at the margins of related projects, and this forces scientists to rely on private financial support from foundations and industry.

The reason for this lack of formal governmental oversight is uncertainty over where to draw lines for permissible research. Religious organizations take a tough stance on chimera research. Most Catholic and Protestant faiths believe God created humans in his image and gave people dominion over the land. They do not believe the line between humans and animals should be crossed.

But when commercial firms respond by pointing out the need to help patients and push the envelope for future medical treatments, political decision makers are paralyzed. Rather than working out a compromise that protects the interests of each side, they have done nothing and trusted industry self-regulation or voluntary guidelines from professional associations to protect societal goals.

As an example, Weissman's work at Stanford on mouse brains is privately financed and subject to oversight from an "informal ethics committee" chaired by a Stanford law professor. Before a biologist undertakes research on human and brain cells, he or she presents a detailed description of his research protocol and the committee offers advice on the propriety of the proposed activity.

However, it is not clear how effective this voluntary oversight would be. When asked about chimeras created between humans and mice, Stanford ethics committee head Hank Greeley said, "There is an ethical concern if we were to confer some sort of human mental ability on a mouse . . . so if you see anything that is not unequivocally mouse behavior, you stop the experiment."[17] Mice have certain neurological mechanisms such as "whisker barrels," which transmit sensory information from their whiskers to brain structures that tell the mouse how to move in the dark. Development of human brain structures in mice with implanted stem cells would represent a checkpoint at which the experiment should be stopped. According to ethicist Greeley, anything that is "not normal mouse behavior" would be grounds for ending the particular biological experiment.[18]

The problem with this reasoning is that it rests on unclear indicators. Overseers measure the ethical line by whether any mouse exhibits unusual or abnormal mouse behavior, or unequivocal human behavior. How would this committee determine whether something was unequivocally mouse-like or evidence of elementary human thinking? The absence of clear thinking on this important question demonstrates the difficulty of self-regulation on chimeras.

In response to these practical problems of oversight, the National Academy of Sciences in 2005 suggested a series of voluntary guidelines for researchers interested in experimental work on chimeras.[19] Among other features, it proposed that no embryonic stem cells be injected into either human or primate blastocysts. This recommendation was designed to avoid confusion between humans and animals. It furthermore said that no animal into which human embryonic stem cells have been placed should be allowed to breed and that research on chimeras should be undertaken only when no other research could be undertaken to provide the information required. Finally, it proposed that "experiments in which there is a possibility that human cells could contribute in a 'major organized way' to the brain of an animal require strong scientific justification."[20]

To oversee this kind of research, the National Academy of Science said that universities and scientific institutes should form official oversight committees that would approve use of human embryonic cells into animals. These committees would be composed of scientists and other university personnel. Scholars interested in undertaking research in this area would need to seek the written approval of this committee and explain exactly what kind of work they proposed to undertake.

Some lawmakers, though, are not satisfied with this approach based on self-regulation. Senator Sam Brownback (R-Kansas) introduced legislation in 2005 that would ban certain kinds of chimera research. Specifically, it would outlaw the injection of human brain cells into animals. In addition, the proposed statute would force the U.S. Patent Office to reject all requests for patents on "chimeras created by combining human and animal embryos."[21] However, so far, the Senate has not acted on this bill. Strong opposition from affected businesses and universities has prevented Congress from even holding a vote on its provisions.

When asked about his response to the Brownback legislation, Weissman strongly condemned the proposed ban. Referring to the Senator, he said, "You better be right, Sam, because what you're doing for sure will affect the lives of people and are you willing to accept the moral responsibility for the lives lost and damage because you didn't like some human cells in a mouse brain?"[22] In Weissman's view, this would be an unacceptable price to pay for possible medical advances in the future.

Unlike the United States, Canada has enacted formal legislation on chimera research. In its 2004 Assisted Human Reproduction Act, it prohibited "transferring a nonhuman cell into a human embryo and putting human cells into a nonhuman embryo." A national government oversight committee was charged with making sure that scientific research is in accordance with these public sector rules. Stiff penalties encourage scientists and businesses to respect these governmental prohibitions.[23]

European Experiences

European nations display considerable diversity in how they approach research on chimeras and other kinds of cross-species research involving human beings. Whereas Great Britain has undertaken chimera research, Germany has banned the creation of animal-human hybrids. This is consistent with each country's approach to biotechnology in general.

Great Britain's approach to chimera research is controlled by the Human Fertilisation and Embryology Authority (HFEA), the same body that regulates most biotech projects. Scientists wanting to undertake projects on chimeras must obtain a license from the government. According to the authority's head of research regulation, "Any research that involves putting a human cell's nucleus into an animal egg would require a license from the HFEA. As with all research involving human embryos, the research team would have to show that the research is both necessary and desirable, and that any embryo created could not be allowed to develop for longer than 14 days or be implanted in a woman."[24]

Following the first human-rabbit embryo created in China, British researchers used rabbit eggs to generate chimeras for the production of human embryonic stem cells. Because of the cost and health risks of obtaining human eggs, scientists argued that these chimeras were a viable way to produce early-stage embryos. Although decision makers called the development of chimeras "a grey area" in terms of the law, scientists agreed to submit their research under the procedures of the HFEA.[25]

Officials from religious organizations object to the procedures being undertaken by researchers in Great Britain. "Research upon embryo 'chimeras' would not only jeopardize the humanity and dignity of the tiny human being, but also hold out a false mirage of hope to some very vulnerable patients," said the director of the advocacy group Movement Against the Cloning of Humans.[26] A spokesperson for the Church of Scotland indicated that "the proposal violated the distinction between humans, who were 'created in God's image', and animals, who were not."[27] When asked about this criticism, Dolly creator Ian Wilmut of Edinburgh University conceded that hybrid embryos were human, but said, "We would only be working at a very early stage with a maximum of 200 cells. I wouldn't think of a human embryo at that stage as being a person."[28]

Figures from government organizations reveal that hundreds of thousands of altered animals now are being used in British laboratory experiments. Overall, animals were employed in 2.8 million experiments in 2005. Of these, 800,000 experiments involved genetically altered animals, some with human genes added. This makes Great Britain one of the leading locales for chimera research around the world.[29]

Research on chimeras is also being undertaken in France, although on a much smaller scale than in Great Britain. After transplanting a number of mouse cells into chicken embryos, French scientists working with British collaborators were able to activate a chicken gene that stimulates teeth growth. Scientists successfully used a chimeric hybrid to create new teeth in this fowl, a dinosaur-era bird whose naturally occurring teeth disappeared 80 million years ago. As a result, it became the only contemporary bird species on Earth to have teeth.[30]

Germany has banned the creation of human-animal hybrids. At the same time that it outlawed cloning, the government prohibited experiments that would combine living animal and human cells.[31] Even xenotransplantation, the placement of animal organs in humans, is controversial. The country has a regulatory authority that oversees work in that area. It requires scientists interested in undertaking work on chimeras to apply for government permission and specify exactly what they plan to do. Germans fear that cross-species research will spread disease and expose human beings to unnecessary health risks.[32]

Despite opposition from some member countries, the European Medicines Evaluation Agency approved the first medical drug manufactured from chimeras. A Massachusetts firm, GTC Biotherapeutics, applied for permission to market a coagulant drug known as antithrombin derived from goats' milk. This product is a result of a human gene added to the embryo shortly after an embryo forms that allows the newly-generated off-spring to produce a blood protein necessary for burn victims, heart attack sufferers, and those at risk of blood clots during surgery. This gives doctors another source of blood protein beyond that donated by humans.[33]

As a sign of how contentious chimeras remain in Europe, when American researcher Norbert Gleicher presented a talk at the annual meeting of the European Society of Human Reproduction and Embryology indicating that he had combined male cells into a female embryo that lasted for six days before it was destroyed, he was widely condemned by his European counterparts. "It is ethically objectionable," said Francoise Shenfield of the society's ethics group. "There are serious scientific and medical problems associated with it. There are potential long-term health risks for any child if this was used in a clinical setting. I cannot conceive of any situation in which this particular technique would be acceptable."[34]

Asian Experiences

Similar to Europe, Asia is divided over chimera research involving humans. China and India are moving ahead in this area, while Japan has been less willing to undertake this type of research. These divisions are comparable to what has happened in other fields of biotechnology in Asia. In many of these areas, China has been a leading innovator, while Japan has expressed more substantial ethical and practical concerns.

In general, cross-species research in Asia is seen as less objectionable than in some Western countries. Many Asian religions do not draw sharp distinctions between humans and animals. Whereas Christian faiths place humans at the center of the universe and cite biblical scriptures giving mankind dominion over animals, Asian philosophies such as Shintoism and Taoism see many animals as having spiritual qualities and occupying more of a level field with human beings.[35]

Some religions in this region (such as Hinduism, the world's third-largest religion) even worship animals such as cows and place them in a privileged position in the universe. Animals are revered because they are seen as receptacles for the human soul. Depending on the kinds of deeds (good or bad) undertaken, humans in a future life may be reincarnated in animal form. This possibility increases the reverence for animals and decreases the significance of the gap between humans and animals.[36]

Like many other examples of biotech innovation, China has few restrictions on chimera research. As noted earlier, its researchers have been in the forefront of work in this area. In the case of the human-rabbit hybrid produced in 2003, the government's ethics advisers placed only one restriction on the chimera that was created. The embryo had to be destroyed no more than 14 days after creation. Chinese public officials did not want the new entity surviving into adulthood.[37]

Hui Zhen Sheng, the Chinese researcher who conducted the experiment, led a research team that was financed by the city of Shanghai. After spending 11 years working at the U.S. National Institutes of Health, she returned to China in 1999 and now is in charge of a 50-person research team. Her goal in merging human and rabbit cells was to develop an alternative option for human eggs in research on therapeutic cloning.[38]

This work generated a host of ethical concerns in the West. Even though nearly all the genetic material in the embryo came from human cells, the use of mitochondrial rabbit DNA provoked fear among ethicists. When asked about the Chinese human-rabbit combination, a spokesperson for the U.S. Conference of Catholic Bishops indicated that "because all the nuclear DNA is human, we'd consider this an organism of the human species." Considering it to be a viable fetus, the church did not support experimentation on or termination of its life.[39]

British pro-life advocates meanwhile proclaimed the chimera "barbaric." Speaking of the Chinese human-rabbit combination, one activist said, "Such abuses of early life, where a new living and genetically unique individual is deliberately created, abused and destroyed with such contempt, must surely send shivers down the spines of the general public. These scientists have claimed that such research is for the good of humankind—yet how is the creation and destruction of such a freak of nature intended to benefit anyone?"[40]

However, none of these ethical issues generated much controversy in China itself. The country's permissive attitude on abortion in particular and stem cell research in general meant that issues of whether the embryos were human, deserved life, or blurred the boundary between humans and animals were not central to public discussions. The fact that Chinese researchers undertook their work unencumbered with opposition from religious forces within their own country liberated them to become aggressive pursuers of chimera research. Indeed, their work has pushed the envelope further than virtually any other place in the world.

Korea, meanwhile, is spending several million dollars a year on xenotransplantation, which is the transfer of living cells from one species to another. The government provides basic funding for this endeavor, and research takes place at universities and in private companies. If sufficient progress is made, the government has promised scientists a fivefold increase in research money over the next few years.[41]

This strategy is consistent with the Korean government's general approach to the marketing of innovative ideas. The public sector likes to target key industries or areas having great economic potential and place a lot of financial resources to develop that area. It followed this path successfully on electronics, semiconductors, computers, cell phones, and (with more limited effectiveness) stem cell research. As scientists make gains and show a payoff from the public investment, the government places more money into that area.

India also has put considerable effort into xenotransplantation. Medical doctors in this country have placed pig hearts in human beings as well as pig spleens and other organs.[42] Fears of "retroviruses," or hidden DNA from past diseases that exist in each species' genes, are less prevalent in India than elsewhere around the world, in part because of the nature of religious beliefs in this country and the close symmetry Indians find between humans and animals. Unlike the fears that have limited the practice of xenotransplantation in a number of Western countries, this possibility is seen as less of a problem in India.[43]

With one-third of the nation living in dire poverty, India has become the "great organ bazaar" for human organ donations.[44] It is the world's largest venue for kidney transplants. Unlike other nations, which have long waiting lists, India provides almost immediate availability for kidneys and other major organs at a low price. This ready supply of human organs has turned India into a major destination for medical tourism.

Japanese law allows the placement of human cells into animal eggs. However, it outlaws the implantation of these animal-human embryos into the wombs either of humans or animals. Those violating the national statute could be given penalties of one year in prison or a fine of 1 million yen (or $9,000). This limits the kind of biotech research that can be undertaken in this area.[45] Much like its attitude on biotech as a whole, Japan has not rushed into much research involving chimeras.

Conclusion

It is clear based on this analysis that many nations have been reluctant to place restrictions on chimera research because of the market potential of this biotechnology and hopes for future medical breakthroughs. With the exception of several Asian nations, such as China and India, most countries are not funding this area with public dollars, at least in terms of human chimeras. Scientists must rely on foundations and private companies to support work in this area.

In places where there is growing pressure to oversee chimera research, such as in the United States, scientists have migrated offshore to Caribbean or Asian sites, such as the island of St. Kitts. They are far from the peering eyes of possible federal regulators or animal rights activists who object to experimentation on primates.[46] Researchers sometimes use non-American collaborators or undertake their work at foreign locations to avoid local oversight.

This example of "country-shopping" illustrates how scientists look for nations with weak regulatory capacity and few opportunities for protesters to disrupt their experiments. For example, Yale biologist Eugene Redmond established the St. Kitts Biomedical Foundation in the Caribbean in 1982. Funded by private donors and foundations, his research team has inserted millions of human brain cells into adult monkeys. Designed to test the possibility that new human neurons producing dopamine would help Parkinson's disease patients who fail to produce sufficient neurochemicals, these experiments alarm some because they produce vervet monkeys with human brain cells.

When asked about the research, Redmond said the reason he undertakes his primate research in the Caribbean is that "primate research has gotten so expensive in the United States one of the other advantages is that it's a beautiful tropical environment so the monkeys can live outdoors and you don't have to provide the air conditioning and heating and all that kind of thing for them."[47]

Additionally, Redmond's team justifies its use of human cells in monkeys by citing the work's potential to help patients with brain disorders. According to Evan Snyder, a neurobiologist who collaborates with Redmond, "Some scientists are completely upset with even a single human cell in a monkey brain. I don't have problems with putting in a large percentage of cells—10 or 20 percent—if I felt it could help a patient. It comes down to what percentage of human cells starts making you squirm."[48]

This is a concern that some have directed at Esmail Zanjani, an Iranian-born hematologist at the University of Nevada at Reno. He has injected sheep with human embryonic stem cells created from bone marrow. His goal was to see if the altered sheep would produce human blood that could fight diseases. He was quite surprised, though, when the stems cells went through the lymphatic system into the liver, heart, blood, brain, and bones of the sheep. In some cases, the added stem cells have led to livers that produce human proteins and are "40 percent humanized."[49]

These are exactly the types of ethical concerns that lead many in the West to condemn chimera research involving humans, primates, and mammals. Given the close genetic relationship between primates and humans (estimated by some as having 99 percent of the same DNA), bioethicists worry that these types of chimeras cross the line beyond what should be considered permissible research. But because some of the research is taking place at locations outside of the United States and in poor places desperately in need of economic development, the country-shopping shields researchers from detailed scrutiny.

Observers look at research on embryonic cell transfers between humans and sheep and wonder whether these processes should be undertaken at all. If embryonic stem cells have the ability to morph into advanced human tissue, what is to prevent the sheep from developing more human-like qualities? This problem is less of an issue for adult stem cells because once they have become livers or kidneys, they will remain that kind of tissue. Yet embryonic stem cells have the capacity to become many different kinds of organs so it is much more difficult to anticipate how their development will affect the chimera being produced.

In its report examining the future of stem cell research with chimeras, the National Academy of Sciences recommended that in order to publish in American scientific journals, authors be required to follow guidelines on chimeras and stem cell research established by the academy. This novel recommendation represents an important attempt to undermine country-shopping by mandating publication standards in prestigious technical journals. By recognizing that scientists can migrate from country to country but still prefer to publish in particular outlets, the academy realized that this gave it a potential power even in an era of globalization. However, so far, there is little evidence that leading scientific journals have implemented this recommendation.[50]

Legal scholars recognize the limit to this kind of approach. Hank Greeley, the Stanford law professor who oversees Weissman's experiments on human brains in mice, stated that "some countries probably won't follow the National Academy of Sciences recommendations or at least some countries may not enforce upon their scientists following those recommendations." He predicted that on chimera research, some countries would end up being more restrictive, while others were far less restrictive in their formal oversight.[51]

This is the reason why it is so difficult for societies to control the science-industrial complex. Scientists realize they have great power due to the economic potential of their research and the mobility of their own knowledge industry. Rather than be dependent on the public sector, these investigators recognize they have the ability to privately finance their research and through advances in communications and technology, they also have the power to move, and therefore escape the regulatory power of the state. This liberates them from societal control and makes it difficult for societies to control chimera experiments that are objectionable.

Chapter 8

Pharmaceutical Companies, Biotechnology, and Health Care

Global pharmaceutical companies play an important role in the health care of consumers and development of new biotechnologies. With 2004 revenues totaling $550 billion, the 50 largest firms produce popular prescription drugs such as Lipitor, Zocor, and Plavix, and sponsor $38.8 billion in research and development. The bulk of pharmaceutical sales come from North America (48 percent of all global sales) and Europe (28 percent), but there also are significant revenues from Japan (11 percent), the rest of Asia (8 percent), Latin America (4 percent), and other regions (1 percent).[1]

"Big Pharma" revenues far outpace the $63.1 billion in revenue generated by publicly traded biotech firms.[2] With long leads on product development and the experimental nature of planned treatments, it has taken biotech businesses many years to reach the point where they are generating sales. However, the industry still has not reached the point of profitability. In 2005, biotech companies lost $2.1 billion, although the positive news for the sector was this loss was down from $4.9 billion the previous year.[3] Since 1976, biotech firms have lost around $100 billion in total.[4]

Although the United States has long been considered the leader in biotech, European and Asian companies are catching up. For example, Europe now has 1,613 biotech firms (up from 584 10 years ago), compared to 1,415 in the United States (up from 1,308 in the last decade). Europe still lags behind American firms in available capital, though.

American companies raised $16 billion in capital in 2005, far outpacing the $3.3 billion raised in Europe.[5]

Because of its financial resources, Big Pharma multinationals have forged strong links with newly emerging biotech companies. With prescription drug revenue growth slowing and key patents on blockbuster drugs expiring over the next few years, pharmaceutical firms need strong connections to the next generation of medical therapies in order to maintain sales. Biotech's scientific expertise and potential for new products provide pharmaceutical companies with an avenue to develop new technologies in return for research investment. The greatest challenges for biotech businesses are obtaining the initial capital to finance expensive personnel and facilities, and gaining the time necessary for new product development. Increasingly, pharmaceutical businesses have put money and licensing agreements in place that allow them to commercialize biotech products.

In this chapter, I examine the funding mechanisms between pharmaceutical and biotech companies. After reviewing the different types of financial linkages and contrasting oversight approaches between the public and private sectors, I outline some of the problems that complicate biotech innovation. For example, critics claim Big Pharma creates strong feelings about particular diseases and then overcharges patients to treat those illnesses.[6] Meanwhile, industry figures complain about patent and intellectual property disputes that limit their ability to develop new products. I close by noting the ramifications of these issues for patients interested in biotech therapies.

The Commercial Funding of Biotech

Biotech remains one of the riskiest investment areas in the financial market. This has been demonstrated over the past few years as the biotech stock market has been a roller coaster, with lots of downs and a few ups. Starting in 2001, many small firms lost 70–90 percent of their stock value. Some have recovered since then, but most companies in this area are well below their historic stock highs.

Yet investors continue to put money into this area because of its long-term potential. In 2003, for example, it was estimated that of the $17 billion raised for biotech companies, more than half ($9.5 billion) came from public stock market funding, $3.3 billion came from private equity investment, $3.7 billion emerged from initial public stock market offerings, and $442 million came from partner payments.[7]

The costs of developing new treatments are enormous. According to Jim Vincent, chairman of the commercial firm Biogen, Inc., "You'd better be prepared to spend 10 years, if you're fortunate, and $500–700 million cumulatively until you start marketing your first drug. You can only raise half of that amount from public markets. That's a serious amount of money for no return over 10 years."[8] According to a Tufts University study, the amount required to introduce a new drug has risen from $231 million in 1991.[9]

In the current period, new drug approvals increasingly are coming from biotech, not pharmaceutical companies. In 2005, for example, 18 of the 29 new drugs approved by the FDA came from small biotech businesses. James Greenwood, the president of the Biotechnology Industry Organization, explained that "the little biotech companies are entrepreneurial. They're risk takers, they will go looking for a brand new molecule to cure a disease that's not been tackled before."[10]

Along the way, though, there are many more failures than successes on product development. Barry Quart, the president of Agouron Research and Development, highlighted the odds against successful treatments when he pointed out that "in the pharmaceutical industry we always realized that we'd be lucky to have 10 percent of good-looking compounds get to market. I've killed off many more products than I've passed."[11] As with any new industry, it is difficult to find money to support untested products having an unclear probability of scientific or financial success.

Of the thousands of biotech companies, very few remain independent. Most have forged some type of collaborative relationships with pharmaceutical businesses. To raise money for product development, biotech firms have developed various kinds of relationships with pharmaceutical companies: partnerships, licensing agreements, and venture capital. These are the ways nascent businesses raise funds to support product innovation. Each is based on differing models of biotech-pharmaceutical company alliances.

Partnerships represent one way for pharmaceutical and biotech organizations to work together. In this situation, companies have a formal relationship to share knowledge, expertise, and product development. For example, a Silicon Valley company called Maxygen developed an alliance with drug company Roche that provided the biotech firm with $230 million in capital investment to work on "next-generation interferon." This included money for research and development, initial fees, and "milestone" fees when performance benchmarks are reached. In return, Maxygen agreed to give Roche rights to commercialize new discoveries and new medical treatments.[12]

One of the virtues of these types of partnerships is they provide up-front money for product development in advance of treatments being approved by government agencies. With the long lead time on drug testing and government approval, these kinds of work relationships provide biotech companies much-needed financial resources at a time when there are no guarantees that scientific discoveries will ever have commercial value. This insulates biotech firms from the ups and downs of the stock market and the varying assessments of investment professionals and their emphasis on quarterly performance.

If a biotech company is completely dependent on stock market investors, the firm's finances follow a "roller coaster" existence. Fortunes rise and fall on the latest news story, product development, and clinical testing programs that succeed or fail. There is little continuity over time and it is difficult for company officials to plan research and development activities, not knowing how investors may react to daily events. Partnerships cushion the volatility of initial stock offerings and public investments by providing a predictable flow of capital funds.

Licensing agreements are another mechanism for collaboration. Under this arrangement, biotech firms consent to licenses that allow pharmaceutical companies to use new advances, drugs, or treatments for commercial development. Typically, these licenses run for a fixed length of time and produce regular payments to the originating company. Unlike partnerships, which represent "across the board" agreements and cover many different inventions, licenses are limited to specific products. Under a license, a biotech firm allows a pharmaceutical enterprise access to particular technologies or drug treatments for a fixed period of time. Companies may license biotech products to many different firms at the same time, in order to finance their firm's research and development.

Licenses offer more flexibility than partnerships. Biotech companies can work with multiple firms at the same time and preserve their autonomy for future product development in other areas. These kinds of arrangements provide much-needed money for biotech product development and give drug firms access to cutting-edge scientific expertise. According to one industry report, licensing deals with biotech firms now account for 70 percent of the pharmaceutical industry's drug pipeline.[13]

Venture capital is a third avenue for biotech and drug organizations to work together. Under this approach, pharmaceutical companies agree to invest capital funds in the biotech business in return for ownership rights. Essentially, companies trade cash and stock ownership. Most biotech organizations lack the experience to deal with drug tests and FDA drug approval. Since pharmaceutical corporations have been doing this for years, venture capital firms allow upstarts to develop and market products.

The downside, of course, is that innovators lose some financial control over their enterprise. Rather than owning 100 (or 80 or 60 percent) percent of their company, they sell stock to private investors to gain access to much-needed capital. Basically, they are trading short-term access to capital from investors for long-term stability. It solves the biotech firm's immediate need for money, while giving outside investors the hope of lucrative profits down the line.

Biotech venture capital is fraught with great risk. There are many things that can go wrong along the path to commercialization: unsuccessful clinical trials, dangerous side effects, and interactions with other drugs. Any one of these obstacles can torpedo a medical therapy and undermine even the most promising new discoveries. If trials do not yield the expected health benefits, it is bad news for partners, licensees, and venture capitalists.

Public versus Private Sector Oversight

Compared to the public sector, commercial firms have looser procedures governing the development of biotech products. According to a National Academy of Science reform proposal for research on embryonic stem cells, there should not be a big difference between the two sectors. Each group should refrain from research on human embryos more than 14 days after gestation. Research should be reviewed by committees composed of scientific experts and the local community. Certain kinds of procedures, such as placing human embryonic stems cells into monkeys, apes, or human embryos, would not be allowed under any circumstance according to the official recommendations.[14]

However, for research funded by government, these guidelines would be mandatory, while in the private sector, they are voluntary. Even if many commercial companies follow the recommendations, there are no guarantees that everyone will follow these suggestions. Pressure to produce profits or commercialize scientific discoveries encourages firms to push the boundary of innovation, even if there are serious ethics questions involved. This freedom allows industry scientists to undertake tests that would not be allowed under public rules.

In California, voter approval of a $3 billion stem cell initiative has unleashed tension between the public and private sectors. Private companies are able to apply for state funding, but insist they will not do so "if too many strings are attached to those dollars."[15] Business leaders complained that the state was attempting to "dictate how much companies can charge for new treatments; maintain the right to patent technology, if companies fail to do so; and require companies to share developments they consider proprietary."[16] Businesses want to maintain intellectual property rights to any discoveries made by their employees, without serious impediment from government agencies.

The conflicting standards between public and private sector procedures for developing biotech illustrate the weakness of the state in the face of the science-industrial complex. Companies do not want to follow the same rules and procedures that are present within the public sector. They want freedom to innovate and a reduction in rules that limit their experimentation. This makes it very difficult to regulate biotech very effectively.[17]

At the state level, companies can play communities off against each other in order to ensure the most favorable rules. If one state, such as California, imposes strict rules on treatment costs or patent procedures, businesses can move to another locale. And if particular countries are too strict, scientists can migrate to other nations where the funding requirements are generous and not so restrictive. This mobility of knowledge aids both large pharmaceutical companies and newly emerging biotech firms. The portability of their work gives them greater power over governments in where and how they undertake research.

The Critique of Big Pharma

The financial power of large pharmaceutical companies has led to vociferous criticisms about "Big Pharma" and fueled scrutiny of the relationship between drug companies and biotech firms. A flood of books with provocative titles such as *The Truth about the Drug Companies: How They Deceive Us and What to Do about It; On the Take: How Medicine's Complicity with Big Business Can Endanger Your Health; Selling Sickness: How the World's Biggest Pharmaceutical Companies are Turning Us All into Patients;* and *Powerful Medicines: The Benefits, Risks, and Costs of Prescription Drugs* has hit the marketplace and focused unfavorable attention on the drug industry.[18]

Adversaries complain that large drug companies inflame people's fears about illnesses, push expensive medications onto the marketplace, spend too much money on marketing, and use political contributions to elicit favorable legislation for themselves. Marcia Angell, for example, says that historically drug companies have generated profits that are 20 to 25 percent of sales revenues, much higher than the 3 percent figure for other industries in the Fortune 500 list.[19] At the same time, marketing costs more than double what these firms spend on research and development. Whereas one-third of sales money goes towards marketing drugs, research expenditures have ranged from 11 to 14 percent.[20]

This has been the case since 1997, the time when the FDA abandoned its previous restrictions on direct company-to-consumer marketing through television advertising. The result has been an explosion of commercials touting health care remedies. In 1998, the first year after deregulation, pharmaceutical firms spent $1.3 billion on advertising, a figure that was 33 times the amount in 1991.[21] By 2005, drug ad expenditures had risen to nearly $4 billion.[22]

The high cost of marketing drives prescription prices up and has led to drug costs in the United States that are 80 percent more expensive than the same medication in Canada and 100 percent more costly than in France or Italy.[23] Through frequent advertising, the pharmaceutical industry targets both general and specific illnesses, and sometimes is accused of manufacturing treatment demands that did not previously exist. According to industry skeptics, this is money that would be better spent on developing useful therapies.

Through campaign contributions and lobbying, the industry has gained favorable rules for itself. In 2004, drug companies donated $17 million to candidates running for federal offices and $158 million on lobbying. There were 18 members of the U.S. House and Senate who received at least $100,000 in contributions. Since 1998, the industry has devoted $758 million on lobbying, which is higher than any other industry in the country. With 1,274 lobbyists, there are two and one-half lobbyists for each Senator and House member.[24]

Although smaller than the drug industry, the biotech industry is devoting more and more money to political advancement. From 1998 to 2002, biotech companies spent $89 million on lobbying political leaders. Over the past decade, they have given more than $13 million in campaign contributions to candidates for federal office. Along with pharmaceuticals, the biotech industry is seeking less regulation, faster drug approval processes, and protection from competitors around the world.

These advocacy efforts appear to be paying off. Unlike most Western countries, for example, the United States does not regulate the cost of prescription drugs. Even when Congress added a prescription drug benefit to Medicare at taxpayer expense, it did not add any restrictions that would keep down the costs of medication. Pharmaceutical interests also have been very successful at preventing the reimportation of cheap drugs from Canada. Even though these medicines have been manufactured by leading companies, drug firms argued that the quality of these imports is suspect.

From an international standpoint, there are complaints that large drug companies do not share medical advances with many developing nations. Diseases are left untreated because of lack of money for prescription drugs. The high cost of medication leaves poor nations and impoverished residents of wealthy countries outside the possibility of obtaining necessary medical drugs.

At the same time, these firms are accused of concentrating their testing on places with few restrictions or consumer safeguards. For example, a book by Sonia Shah entitled *The Body Hunters* explores how "the drug industry tests its products on the world's poorest patients." Among other claims, she argues that pharmaceutical companies have shifted clinical work to poor nations in order to escape stringent drug testing regulations in the West.[25]

These claims are consistent with the "country-shopping" argument of earlier chapters. If members of Big Pharma and the science-industrial complex do not like the regulations of particular nations, they shift their activities elsewhere with fewer restrictions. According to Shah, this sometimes has led to tragic results in terms of human death and illness.[26] Rather than experiment according to accepted Western standards, companies have tested drugs on desperate individuals who lack the resources to protect their health and well-being. This raises serious ethical concerns about the propriety of some kinds of clinical trials.

Patent and Intellectual Property Disputes

While critics attack pharmaceutical companies for excessive profits and unfair practices, the biotech industry complains that changes in its legal environment have undermined its ability to innovate. Uncertainties involving patent and intellectual property issues have complicated product development and led to costly lawsuits. Disputes over who developed what products and who should benefit financially plague biotechnology and impede advancement of the industry.

Much of the basis for these intellectual property conflicts date back several decades to a landmark court case. In 1980, the U.S. Supreme Court considered whether a microbiologist could patent an organism able to clean up petroleum spills by degrading oil products. On a 5–4 vote, justices ruled that scientist Ananda Chakrabarty could do so, and this decision ushered in an era of patents for a wide range of natural products in the United States.

The case altered the scientific and financial environment for all American inventors. Items from living tissue, cell lines, microorganisms, and genetically modified plants subsequently have won legal protection. This has allowed their creators to treat discoveries as intellectual property and charge those who wish to use them for commercial purposes. Since then, more than 20,000 patents relating to genes have been filed with the U.S. Patent Office.[27]

Over time, similar rules were adopted in other countries. The new patent laws produced an explosion of scientific creativity and product commercialization. In Europe, between 1991 and 2000, for example, the number of total patents issued grew by 70 percent, from 58,000 to 98,000. This growth reflected rising interest in the commercial application of scientific research in various nations.

Befitting the growing power of the European Union, patents on the continent have shifted away from the country-by-country model of earlier eras. Rather than seek patent protection in Germany or France alone, nearly all the current patent applications are for Europe as a whole through the European Patent Office.[28] This has allowed European discoveries to diffuse quickly throughout the continent and get to needy patients.[29]

However, the United States has broader patent rules than virtually any nation around the world. Unlike European countries, which are "process-driven" and cover procedures to make things, the American approach is "product-driven" and applies to specific creations or discoveries. For example, according to U.S. law, a patent can be obtained for "any new and useful process, machine, manufacture, or composition of matter, or any new and useful improvement thereof."[30] This has allowed inventors to patent genes, genetic markers, or living materials that are not covered in other lands.

Patents are a key ingredient in scientific innovation in many nations. Because they allow creators to benefit financially from important discoveries and control how that product is commercialized, these kinds of legal protections provide substantial incentives for developing new knowledge. Companies cannot develop distinctive products without authorization from those who created the item. In the same way that authors or music producers gain royalty income, patent and licensing fees represent a key fuel for the science-industrial complex.

Generally, patents between governments and inventors last for a specified period of time (generally 20 years). During that period, the scientist can license his discovery for use by others in return for payments. Following expiration of the patent, others can make use of the discovery without paying fees to the inventor. This allows scientific discoveries to be used by a wide range of people.

Financial incentives offered by patent laws represent a major boon to the biotech industry. It is not likely that the area would have flourished to the extent that it has without these types of legal mechanism. If there were no way to safeguard discoveries or benefit materially from commercialization, we would not have so many start-up companies seeking to develop new knowledge.

At the same time, if scientists discover that someone has infringed on their patent and is using the discovery without licensing, inventors can sue and win up to three times the amount of provable damage. This provides teeth to enforce the sanctity of scientific creations. Companies must respect the ownership of ideas and patent infringement lawsuits have become a big business in many locales.

The downside of these legal protections, though, is that although biotech offers considerable future promise, patent and intellectual property disputes represent major barriers to industry development. With the sharing of information by scientists across national boundaries and differing patent laws between countries, there are many grey areas of patent infringement that become the object of lawsuits and contentiousness. Some biotech companies spend 5 to 10 percent of their budget on legal fees.[31] This is a major challenge for emerging firms attempting to develop new products.

Lawsuits represent a particular problem in the biotech area because nearly all of that business is based on intellectual property. Especially in the early days of a company, when there are no specific products to sell, raising capital depends almost entirely on perceptions about scientific discovery and the possibility that certain inventions have commercial value. It is hard to generate capital when patents are unclear or competing companies sue on grounds of patent infringement.

There are big differences across countries in what can be patented and for how long.[32] In Germany and Japan, for example, it used to be that only methods for making pharmaceuticals were patentable, not the compounds themselves. But because this narrow interpretation was seen as restricting incentives for scientific discovery and commercialization, both countries changed their laws to move more in accordance with the rest of the world.[33]

In 2005, India agreed to change its laws toward the standards set by the World Trade Organization's agreement on Trade-Related Aspects of Intellectual Property Rights. The goal of this agreement was to "create a harmonized global system under which inventions are granted exclusive marketing rights for a minimum of 20 years for 'new and inventive' products."[34] Prior to this treaty, 50 developing nations around the world excluded pharmaceuticals from patent protection on grounds that only specific production processes should be safeguarded. This allowed those countries to market drug "knock-offs" designed to mimic pharmaceutical prescriptions but at a much lower price than those sold by Big Pharma. India's biotech industry had sales of $370 million in 2003. The sale of generic drugs constitutes around 80 percent of biotech revenues in that country.

When China joined the World Trade Organization in 2001, it extended patent laws to 20 years and created mechanisms for compulsory licensing. That country now has a $29 billion pharmaceutical industry and a biotech sector with $50 million in sales. However, most of the revenues from both industries come from copying Western drugs.[35]

There is some disagreement over the impact of patent laws on innovation. One paper looked at Japanese patent reforms in 1988 and found that the extended scope of patents had a "quite modest" impact on scientific discoveries. Rather than becoming a boon to new creation, the longer patents merely offered greater protection for existing products.[36] However, other research shows stronger effects. An analysis of international product innovation found that stronger intellectual property rights protection "increases the rate of product innovation [and] product transfer."[37]

The imperfect relationship between patent law and innovation demonstrates difficulties in the ability of governments to regulate biotech innovation. With no single global authority and a mishmash of rules and procedures across countries, it is difficult for companies to safeguard their commercial interests. Firms that do not want to deal with strict patent rules in the industrialized world can shop for nations that are not members of the World Trade Organization or have not ratified stringent intellectual property rules. This allows them to make generic drugs, market knockoffs, or infringe on the patented work of major inventors. It is a major challenge for biotech firms to protect their discoveries across the entire world.

Ramifications of Biotech Health Care For Patients

Fifty years ago, the development of penicillin and other antibiotics allowed doctors to treat infections and save people's lives. These "miracle" drugs were primary weapons in the arsenal to improve health care. Through these and other advances, scientists improved the quality of health care and saved many lives in the process.

Today, though, drugs have become very specialized and are designed to treat particular symptoms. Effectiveness varies considerably with the particular drug and the individual. Some cancer drugs, for example, have an effectiveness level of only 20 percent. Success rates for major drugs as a whole are estimated to range from 40 to 75 percent.[38]

At the same time, many prescriptions come with side effects that sometimes seem worse than the illness being treated. A treatment for cancer, for example, may cause hair to fall out. Cholesterol-lowering medication may lead to rare liver ailments. Headaches may accompany some prescription drugs.

These side effects create a serious dilemma for patient care. Should medical personnel prescribe drugs to patients knowing the treatment will be ineffective for significant numbers or carry side effects for others? Many doctors right now employ trial and error approaches to treatment because they do not know what medications will work with which patients. If one drug does not work, they try something else until they hit on an effective remedy.

What is needed, of course, are ways to personalize medicine so that prescription drugs can be targeted on those most likely to benefit. This is exactly the hope of biotechnology, that treatments can be tailored to the genetic structure of particular individuals. This era of personalized medicine offers the prospect that better matching of treatment and patient will reduce side effects and increase the efficacy of particular medications.

One chip called an AmpliChip measures 30 different variations on genes controlling the speed with which the liver metabolizes drugs. Some people have a rapid metabolism that integrates the drug quickly, leaving it little time to affect the patient. Others ingest material slowly, which increases the odds of side effects. By knowing someone's genetic composition, doctors can determine which drugs are likely to be effective with particular individuals. They then could tailor treatment programs with far greater precision.[39]

Biotech-based medicine furthermore offers the hope of remedies that focus on certain racial or ethnic groups. For example, African Americans are at greater risk than other groups to diabetes, high blood pressure, and sickle cell diseases. By understanding how different groups or men versus women react to medicine, it advances the ability of health care professionals to treat specific illnesses.

Currently, there are 230 treatments based on biotechnology that already have been approved by the FDA. This includes remedies for cancer, insomnia, multiple sclerosis, and severe pain. There are another 250 drugs in late-stage clinical trials that are awaiting FDA approval. Within a year, global sales of biotech products are expected to rise to $8 billion.[40]

But for treatments to be developed, there need to be clinical trials. In the United States and Europe, there are stringent rules on patient selection, informed consent, and protection against harm. Medical personnel cannot test products on human beings without informing them of particular risks and getting their written consent. The goal is to make sure humans are not exploited unfairly or unsafely.

In places such as India, China, and Mexico, though, there are few restrictions on human clinical trials. Scientists can inject experimental drugs or use unreliable medical procedures without public oversight. Rules on informed consent are lax and sporadic. There are few guarantees that patient safety is protected very effectively.

With its patent reforms, India is becoming a big place for clinical trials. That nation has the virtue of low cost, a ready supply of people willing to participate in trials, and a large number of individuals who have not previously been treated with other medications. This lack of prior pharmaceutical experience is ideal for drug testing because it makes for a purer assessment of drug impact.[41]

Developing nations also have become important locales for medical tourism. These are places where visitors can seek treatment for experimental therapies not legal in the West or get human organs without being placed on a long waiting list. China, India, and Thailand are nations where medical tourism is popular. There are short lines for organ transplants. And experimental stem cell treatments are offered.

There is controversy over some of these offerings. In the United States, there are strict rules governing organ transplants. Favoritism is not allowed, and doctors require reasonable health and certain age ranges to qualify for a transplant. In China, in contrast, some have complained that human organs are harvested from death sentence prisoners shortly after execution.[42] In 2005, for example, China executed 1,770 prisoners, far more than the 60 in the United States. Mobile vans in China now administer legal injections, and organs are cut out of the dead corpse immediately upon administration of the capital punishment. Amnesty International claims that "we have gathered strong evidence suggesting the involvement of (Chinese) police, courts and hospitals in the organ trade."[43] Critics say consent rules in China are weak and do not protect the donor.

Currently, there are no rules on Americans getting treatment in other countries. People can get treatments forbidden at home as long as they are willing to suffer personal risks. Coretta Scott King, for example, died seeking treatment for ovarian cancer at an alternative Mexican clinic. Told by American doctors in November 2005 that she was terminally ill, Mrs. King traveled to Mexico to obtain remedies that were unavailable in the United States.[44] She passed away before she even could start her therapy. Her medical experience reveals the extent to which people are willing to tolerate health risks from experimental therapies when their lives are in serious jeopardy.

Chapter 9

Whose Ethical Standards?

In this book, I argue that a biotech science-industrial complex has emerged in many places around the world. Based on close collaboration between commercial enterprises and scientists at leading universities and research institutes, this development is global in nature and independent of state control. With increasing collaboration across national boundaries, biotech globalization differs from economic-based globalization. Rather than focusing mainly on trade liberalization and financial transactions, this new biotech globalization involves plant modification, genetic manipulation, and human reproduction and raises a range of ethical considerations in the policymaking process.

There are a number of features that have brought science and industry closer together, many of which have also weakened the power of political institutions: deregulation of the public sector, corporate partnerships with colleges and universities, "country-shopping," "scientist-buying," medical tourism, and the prevalence of contraband goods in biotech trade. These developments have made it more difficult to regulate biotechnology and have undermined the power and autonomy of the state.

Using a cross-national perspective focusing on the United States, Great Britain, Germany, France, China, Japan, India, and Korea, I examine five cases of biotechnology: in vitro fertilization, genetically modified foods, cloning, stem cell research, and chimeras. In looking at how these nations have handled these policy areas, it is clear that each has different mechanisms for deciding biotech controversies. Some countries value scientific expertise while others weigh religion and public opinion more heavily. In some places, political institutions are permeable, while elsewhere the state has more capacity and independence. In still other locales, economic interests and the size of the biotechnology industry shape policy choices, while in other places these factors do not matter very much.

The result is considerable variation in how countries manage biotech issues. Depending on the particular configuration of religious, political, and economic forces, nations range from more permissive to more restrictive in what they allow. The United States, for example, is permissive on in vitro fertilization but stringent on cloning and stem cell research. Great Britain has among the most permissive rules on nearly every type of biotechnology. Conversely, France and Germany tend to be stringent on all types of biotech. Within Asia, the countries of China, India, South Korea, and Singapore are permissive, while Japan generally is restrictive on most fields of biotechnology. Other than the United States, which exhibits schizophrenia in its biotech policies, countries that are permissive on one field of biotechnology generally follow the same approach toward other areas.

In this chapter, I step back from the particular features of policymaking in these countries and examine the question of who should decide policy issues involving fundamental ethical considerations. Whose values should play a role in biotechnology—those of religion, politics, or economics? Should companies, scientists, universities, public regulators, elected officials, or the general public control these matters? If there is a place for voters and politicians in biotechnology, how should their views be incorporated in decisions that are made?

Who Does Decide? IPE versus RPE

There are two models of decision making on biotechnology. Nations that tend toward biotech permissiveness generally do so for reasons of international political economy (IPE). These are nations with either budding or established biotech companies whose political lobbying makes governments sympathetic to deregulation or reliance on industry self-regulation, or professional guidelines for societal oversight. Religion here is not a major opposition force to biotechnology, and public opinion is sympathetic to the possibility of medical advances. In these situations, the government is content to allow the science-industrial complex to operate as it sees fit, subject to disclosure requirements or weak regulation.

In contrast, countries that are more restrictive on biotechnology typically represent cases of religious political economy (RPE). These places have stronger religious or cultural organizations that intervene against liberal biotech policies and that teach people that certain procedures are ethically suspect. Cultural values in these areas tend to be conservative and traditional, and there are suspicions that the science-industrial complex is not serving broader social interests. Often times in these countries, the biotech sector is weak and does not have a lot of economic or political power. This allows religious and cultural forces to gain power over economic interests and to place restrictions on biotech innovation.

Of course, as already mentioned, the United States is a special case of schizophrenic policymaking on biotechnology. Depending on the biotech area, it has elements both of IPE (strong biotech, powerful agribusiness, and sympathetic public opinion) and RPE (strong religious organizations and unfavorable public opinion). This moves the country in a permissive direction on in vitro fertilization and genetically modified food (and sometimes on chimera research), but in a more restrictive direction in regard to cloning and stem cell research.

In looking at particular nations, political institutions in some places have been able to carve out autonomy for themselves and take a firm stance in favor of biotech regulation. For example, Germany, France, and Japan have enacted legislation that places strict controls on many types of biotechnology, especially genetically modified food, cloning, and stem cell research. Italy, Germany, and Ireland have stringent rules on in vitro fertilization and assisted reproduction. The United States has regulatory mechanisms in place to control some of the research that takes place in regard to cloning and stem cell research.

When nations place restrictions on biotechnology, it typically is due to a combination of religious and cultural forces that allow these countries to use social pressure as a mechanism for strong state capacity. In these cases, moral concerns lead the public sector to overcome lobbying from the science-industrial complex. As political conflict moves from the technocratic area of low visibility and low public conflict to partisan and ideological conflict involving fundamental values, the greater visibility and salience of these subjects allows voters and politicians to place restraints on biotech innovation.

In examining five cases of biotechnology, though, the most common political reaction is inaction, weak national laws, reliance on industry or professional self-regulation, or ineffective formal regulation. Strong economic interests in higher education and business push for leniency in the rules of the game. Using the permeability of political institutions, they argue that biotechnology is good for trade and economic development. They tell the public that scientific progress and innovation is crucial to future well-being. When the public is supportive of biotechnology and economic interests are in favor of experimentation, it is hard to build coalitions restricting biotechnology.

Even in countries where stringent laws have been passed, effective regulation remains difficult. In an era of globalization, scientific knowledge is mobile. Economic interests play countries off against one another. If one nation passes restrictive laws on certain procedures, scientists and business "shop" for other countries that are more lenient. Contraband trade in biotech products continues despite efforts to outlaw those particular products. This has especially been true in the case of genetically modified food. The globalization of the science-industrial complex has weakened the state and made it difficult to regulate experimentation thought to be detrimental to the public interest.

The result is that in many cases, voters and elected officials have been marginalized in their ability to bring societal values to bear on biotech decisions.[1] In some cases, decisions have been privatized to companies and universities. And in other cases, the science-industrial complex bypasses procedures for regulation by migrating to places where rules are nonexistent or enforcement is lax. In either scenario, it is hard for collective values to limit the kinds of experiments and commercialization that currently are taking place.

Who Should Decide?

The weakness of political institutions and the ability of economic interests to play nations against one another does not mean that the public should have no role in biotech policymaking.[2] Given the fundamental nature of genetic engineering and the ramifications for human life and dignity, citizens have a right and a responsibility to bring societal values into the policymaking process.[3] Biotech raises important questions for entire societies, and political institutions need some way to balance the competing objectives of science, industry, and morality.

It is insufficient for scientists to argue they know better because of their specialized knowledge or for industry to make grand claims about trade and economic development. Neither technical training nor profitable enterprises guarantee that wise policies are produced. Flawed decisions result when ethical considerations are pushed to the sidelines.

As pointed out by ethicist Michele Curtis, science and religion occupy separate worlds with their own vocabulary, assumptions, and ways of thinking.[4] Science focuses on the principles of objectivism, positivism, and reductionism. Truth is broken down into component parts, tested, and certified in a factual manner. Hypotheses are evaluated and knowledge accumulated over time.

In contrast, religion occupies the realm of holistic thinking and faith in basic principles. People apply abstract ethical principles to concrete cases and object when lines of morality are crossed. They look at situations as a whole and draw inferences about the decline of civilization or threats to human dignity. They are intense when fundamental beliefs and values are threatened.

In reality, neither worldview is as pure as it likes to present itself. With the collaboration of science and industry, scientists no longer are truth-seekers unswayed by profit motives or commercial appeal. They work on projects deemed potentially profitable, and exhibit a growing number of conflicts of interest between their sponsors and their scientific projects.

In the same vein, religion does not always exist on high and principled planes. Religious authorities stereotype opponents as "murderers" and fail to keep up with scientific advances that mandate changes in philosophy. Sometimes they border on being Luddites who oppose any new innovation simply because it previously has been opposed by someone in the religion's hierarchy.

On the crucial issue of the beginning of life, for example, it is instructive to note the major differences in opinion on when life starts across the major religions. In biotechnology, these differences in beliefs and assumptions often spell the difference between favoring and opposing particular procedures. For example, Christians say life begins at conception. However, many scientists favor the 14-day mark when rudimentary skeletons and human organs appear in the fetus. And Moslems and Jews favor a 40-day threshold on the grounds that this is when the soul enters the fertilized egg.

Clearly in the area of cloning and stem cell research, these variations in views about the initiation of life hold major ramifications in deciding when it is appropriate or acceptable to alter human embryos. If one thinks life begins at conception, no experimentation whatsoever should be allowed. However, if the threshold is set at 14 or 40 days into the development of the fetus, that allows many more possibilities for human intervention.

Given the wide range of differences in beliefs between fundamentalists and secularists; Protestants, Catholics, Jews, and Moslems; and Eastern versus Western religions, it is not likely there ever is going to be consensus on these crucial points. People of different religious and ethical backgrounds are going to disagree sharply on the application and resolution of these matters. The only way things will get resolved is through improvements in the mechanisms of deliberation. Religious, moral, and ethical viewpoints need to be incorporated in biotechnology, and public discussions must move beyond standard stereotypes to a deeper consideration of controversial issues.

Legislatures versus Bureaucracies?

The big question in thinking about how to resolve complex moral controversies is the degree to which decisions should be made by legislatures versus delegating authority to unelected bureaucrats, experts, and administrators. Scholar Timothy Caulfield, for example, prefers the delegation model. Citing the case of Great Britain, where parliament enacts broad principles but leaves the sorting out of regulations and procedures to advisory councils and government administrators, he believes that "recommendations for governance should be broadly and generally framed to allow the regulatory scheme to adapt to changes in science and social mores."[5] According to his perspective, the details of regulation should be left to regulatory agencies. These are the individuals who have the knowledge and expertise to decide matters of policy and who have the flexibility to adapt as new circumstances warrant.

While there may be some virtue in that approach, Caulfield underestimates the danger that regulatory bodies may be captured by interested parties. There is lengthy literature suggesting that over time, outside economic interests take control of administrative agencies and that a "revolving door" between industry and government undermines the ability of agencies to exercise independence over the regulated industry.[6] Given the power of the science-industrial complex in many countries, it is difficult to assume that administrators can retain the independence needed for oversight of biotechnology.[7]

The problem in some biotech areas is industry concentration. For example, a handful of agribusiness companies control the area of genetically modified foods. These multinational corporations exercise tremendous power over how products are modified and how they are distributed around the world. Furthermore, it has been estimated that just five biotech companies now control more than 95 percent of gene transfer patents.[8] This control gives them enormous leverage over governments and marketplaces.

Given the industry concentration of economic power and the close collaboration between science and industry, who can speak for consumers, farmers, citizens, and those opposed to biotech innovation? If past experience is any indication, it is not likely that these politically weak people will receive much representation through administrative bodies. It is far more likely that their views will be heard through legislatures, since elected officials need their votes in order to win elections.

This suggests that the administrative delegation model is inadequate for biotech decision making. For consumer voices to be heard, decisions must involve elected officials and extensive media coverage. Increasing the visibility of conflict in this area is necessary for nonindustry perspectives to be represented in the policy discussion. That is the only way to broaden the discussion to noneconomic interests.

National versus International Regulation?

With the globalization of the science-industrial complex and the prevalence of country-shopping, it is clear more and more that regulation needs to move to the international level, beyond the activities of individual countries. It is very difficult for specific nations to regulate multinational corporations or to police international collaborations. The mobility of knowledge makes it virtually impossible that such efforts will be successful.

It is instead mandatory that international bodies such as the European Union, the World Bank, the Agency for International Development, and the United Nations become more active in overseeing biotechnology. These institutions have the regional or global reach that gives them more capacity for effective oversight. Through united action, these bodies can harmonize regulatory approaches and reduce the fragmentation in oversight activities.[9]

Increasingly, the European Commission is attempting to forge consensus across nations on the continent regarding biotechnology. Although such an attempt has been difficult given the religious, political, economic, and cultural differences across Europe, the Commission has organized conferences that have contributed to international agreements as well as undertaken policy changes that are more unified than in the past.

For example, the European Union passed the Novel Food Regulations Act of 1997 that harmonized the approval process and labeling requirements of genetically modified foods in Europe.[10] This allowed all countries on the continent to have a more consistent approach to this policy area. Such efforts bode well for the ability of international organizations to counterbalance the science-industrial complex and the power of multinational corporations.

The European Commission has also created the European Group on Ethics in Science and New Technologies. Its purpose is to provide support for bioethics research and elevate the place of ethics in biotech policymaking. These kinds of activities make it possible for opinion leaders and decision makers in various countries to learn from one another and determine what the best course of future action should be.[11]

In the end, the only way to control the globalization of biotech innovation is through global institutions. Without the economy of scale provided by global organizations, it will be virtually impossible for societies to safeguard their legitimate interests in biotech policymaking. International organizations have the necessary clout and the capacity to represent more collective interests in the political process. They can help create viable mechanisms for public oversight of new advances in biotechnology.

The Need for Public Education

The most pressing concern in biotechnology is the need for greater public education. The current level of citizen knowledge is low. Public opinion surveys in many nations reveal that ordinary citizens lack fundamental knowledge of science and technology.[12] They have difficulty answering basic science questions correctly, and do not understand the fundamental mechanisms of human genetics.

In the United States, for example, there is very little understanding of what cloning or stem cell research actually involves and how it is undertaken. Polls furthermore demonstrate that most people think they are not eating genetically modified food even though most clearly are. It is hard to argue citizens deserve more of a voice in public policymaking when they know little about scientific matters.[13]

Every two years, the National Science Board in the United States conducts a survey of public opinion about science and technology. The good news is that literacy is edging up. Knowledge of basic science has improved over the past decade. However, the bad news is that there remain fundamental misperceptions of basic knowledge. Only 51 percent of Americans correctly noted that antibiotics do not kill viruses, only bacteria. This was up from 40 percent a decade ago. Just 45 percent were able to accurately define DNA, the fundamental building block of life.[14]

Studies of European Union nations reveal a similar problem. For example, a true/false quiz of European residents found that 43 percent erroneously think that antibiotics kill viruses as well as bacteria, 29 percent incorrectly believe that the sun revolves around the earth, and 20 percent inaccurately claim that it is a mother's genes that determine whether a baby is a boy or girl.[15]

There is variation across European countries, with Sweden having the highest percentage of correct answers (79 percent), while Turkey (44 percent correct), Bulgaria (48 percent accurate), and Cyprus (49 percent right) have the lowest scores. There was also a gap between men and women with men answering 70 percent of the questions correctly, compared to 62 percent for women. Attending religious services also made a difference. Seventy percent of those who never attended church provided correct answers, while only 54 percent of those who regularly attended made accurate responses.[16]

Yet while the public's knowledge of basic science is rather low, support for pseudoscientific approaches remains substantial. In the United States, for example, more than one-third of the population believes in astrology, i.e., the pseudoscience that believes planets and stars shape human lives. Nine percent feel it is very scientific and 31 percent believe it is somewhat scientific. In Europe, even larger numbers believe in astrology, with 53 percent of respondents who claim astrology is "rather scientific." Beyond astrology, half of Americans believe in extrasensory perception and 30 percent agree that "some of the unidentified flying objects that have been reported are really space vehicles from other civilizations."[17]

In Japan, citizens have even lower levels of knowledge about science than in Europe or the United States. Only 23 percent correctly answered the question about antibiotics not killing viruses. Only 25 percent rightly said that a father's genes determine the gender of a couple's baby.[18] The low knowledge of science and technology in Japan and other Asian countries is a challenge for policymakers wanting to push ahead on biotechnology.

Even when polls indicate majority preferences on a particular issue, it is obvious opinions are not very deep or well thought out. Small differences in question wording make a substantial difference in how citizens respond to the policy option. This certainly is true in the case of cloning. As an illustration, preferences can move nearly 20 percentage points depending on whether cloning is being undertaken for "abnormalities in embryos," "infertility treatment," or "to produce copies of humans for organs to save others."[19]

A variety of improvements obviously are needed to make the public more informed about biotechnology.[20] There need to be forums for reasonable discussion about the goals and objectives of work in this area. Public hearings should be held when major biotech legislation or regulations are considered. When new rules are considered by authorities, there needs to be a mechanism for public participation in those decisions.

Universities and other educational institutions need to take their responsibilities seriously. Attention must be devoted to understanding the societal and ethical ramifications of biotech research. Right now, there is little discussion of the politics or sociology of biotech and this perpetuates public ignorance about biotechnology.

In addition, more attention needs to be devoted to news coverage. According to U.S. and European surveys, the vast majority of citizens learn about science from television more than newspapers, museums, or the Internet. For example 53 percent of Americans cite television as their major source of news, while in Europe, the figure naming television as the top source rises to 60 percent.[21]

The mass media in nearly every country should devote greater attention to coverage of biotech issues. Currently, the press covers biotechnology only when there is a new creation (such as Dolly the sheep), powerful pictures, or a major scandal (such as when a South Korean scientist fabricated stem cell results). While these developments are important, there should be more sustained coverage of scientific breakthroughs, the economics of science, science-industry collaborations, and government policy on science and technology.

The only way for the public to become more informed about these matters is through serious coverage of difficult issues. Matters such as education and health care receive far more attention than does biotechnology. This weakens the opportunities for public education and keeps citizens from learning more about issues that affect society and themselves. Unless biotech garners more news coverage, it will be difficult for citizens to have the information resources necessary for informed decision making.

Some regions of the world suffer from undue cynicism. For example, Europeans show considerable skepticism about science in general. When the European Commission in 2001 asked 16,000 people their views about science and technology, 43 percent agreed that "scientists are responsible for the misuse of their discoveries by others." Eighty percent felt legal authorities needed to "enforce formal ethical rules for scientists." Deep cynicism about science pervaded even among the most educated members of the sample. Fifty-nine percent worried that transgenic crops might harm the environment, while 95 percent want to choose whether they purchase genetically altered or "natural" food products.[22]

Biotech Globalization and the Risk of Backlash

One of the ironies of the contemporary period is that biotech globalization seeks tighter control over technology at the very time such innovation is decentralized, fragmented, and under the control of nonstate actors. This represents a combination that is frustrating to moralistic actors because they cannot control the pace of innovation, yet also frustrating to scientists who resent state interference in their scientific activities.

In the short run, it is not likely that nation-states are going to gain much greater power over the science-industrial complex. Biotechnology is an area of relatively weak state authority. With low transaction costs and transferability of knowledge, it is not clear how governments can become more powerful in controlling innovation in the immediate future.

In the long run, though, the anxiety that people feel about new technology and frustration over the inability to exercise much control creates the risk of public backlash that could be very problematic for the future of innovation. Civil or religious forces could limit the ability of scientists to innovate or force the state to exercise much stronger authority over technology policy. Fundamentalist forces in various countries are worried about the sanctity of life, and they are pressing for outright bans or prohibitions of various kinds of procedures.

If this development gains strength, it would move biotechnology policy much closer to the politicized and ideological battles that currently take place in the areas of education, health care, and social welfare. Rather than being a depoliticized policy area controlled by technical experts, biotechnology policy would become more high profile, conflictual, and emotional in its decision making. In the right set of circumstances, religious movements could gain power and use political institutions much more effectively than has happened in the past.

In addition, there is the risk that biotech experimentation will result in disaster. Right now, Chinese and Indian researchers are undertaking clinical trials on humans through the use of stem cells. There is virtually no oversight of these activities and little understanding of possible ramifications produced by injecting stem cells directly into patients. In this situation, the risk of damage is quite high, as unregulated experimentation on human beings is a recipe for biotech disaster. It shows how country-shopping by scientists and patients could lead to catastrophic outcomes, if not controlled by proper authorities.

Finally, there is the possibility that biotech advances will be directed toward ignoble callings rather than noble ones. Sometimes, for example, past scientific inventions have been employed for purposes of military conquest, national aggrandizement, or pure profiteering. Rather than serving human society, new creations have been harnessed either by the public or private sectors for selfish or evil ends. This limits the benefits of science and technology, and redirects human energy away from its desired benefits. Unless technology is harnessed toward socially desirable goals, people will not feel comfortable with the expansion of globalization and the transfer of new technology.

However the political dynamics of state control turn out, it is important to understand the new biotech globalization. More and more cases of new technology involve ethical issues and it is crucial to determine how these examples differ from the economic globalization of past years. By studying the ethical dimensions of technology transfers across national boundaries, research can shed light on the factors that determine policymaking on contentious, moral issues.

Notes

Chapter 1

1. Organisation for Economic Co-Operation and Development, *Compendium of Patent Statistics* (Paris: OECD, 2004), 16–18.
2. See information available at www.ScienceWatch.com.
3. National Science Board, *Science and Engineering Indicators 2004* (Washington, D.C.: National Science Foundation, 2004), 0–6.
4. Paul Kimball, "Globalization, the U.S. Economy & The Imperative of Innovation," (presentation, Brown University, RI, March 16, 2005).
5. Jan Fagerberg, "Technology and International Differences in Growth Rates," *Journal of Economic Literature* 32, no. 3, (1994): 1147–1175.
6. Hugh Slotten, "Satellite Communications, Globalization, and the Cold War," *Technology and Culture* 43 (April 2002): 315–350.
7. Organisation for Economic Co-Operation and Development, *Science and Technology Statistical Compendium* (Paris: OECD, 2004).
8. Ibid.
9. National Science Board, *Science and Engineering Indicators 2004* (Washington, D.C.: National Science Foundation, 2004), 0–16.
10. Organisation for Economic Co-Operation and Development, *Science and Technology Statistical Compendium* (Paris: OECD, 2004).
11. National Science Board, *Science and Engineering Indicators 2004* (Washington, D.C.: National Science Foundation, 2004), 0–4.
12. Ibid., 4–52.
13. Derek Bok, *Universities in the Marketplace: The Commercialization of Higher Education* (Princeton, NJ: Princeton University Press, 2003).
14. Jennifer Washburn, *University, Inc.: The Corporate Corruption of Higher Education* (New York: Basic Books, 2005).
15. Arti Rai and Rebecca Eisenberg, "Bayh-Dole Reform and the Progress of Biomedicine," *American Scientist* 91 (2003): 53.
16. Melody Petersen, "A Conversation with Sheldon Krimsky: Uncoupling Campus and Company," *New York Times*, September 23, 2003, F2.
17. Sheldon Krimsky, *Science in the Private Interest* (Lanham, MD: Rowman & Littlefield, 2003).
18. Goldie Blumenstyk, "Colleges Cash In on Commercial Activity," *Chronicle of Higher Education*, December 2, 2005, A25–26.

19. Goldie Blumenstyk, "Turning Research—Slowly—Into Riches: Technology Transfer Gains a Foothold at European Universities," *Chronicle of Higher Education,* October 7, 2005, A44–45.
20. Daniel Drezner, "The Global Governance of the Internet: Bringing the State Back In," *Political Science Quarterly* 119 (Fall 2004): 477–498; and Daniel Drezner, *All Politics is Local* (Princeton, NJ: Princeton University Press, 2007).
21. Robert Wade, *Governing the Market: Economic Theory and the Role of Government in East Asian Industrialization* (Princeton, NJ: Princeton University Press, 1990).
22. Atul Kohli, *Democracy and Discontent: India's Growing Crisis of Governability* (New York: Cambridge University Press, 1990).
23. Peter Haas, "Introduction: Epistemic Communities and International Policy Coordination," *International Organization* 46 (Spring 1992): 1–35.
24. AnnaLee Saxenian, *Regional Advantage: Culture and Competition in Silicon Valley and Route 128* (Cambridge, MA: Harvard University Press, 1994).
25. Paul Wapner, "Politics Beyond the State: Environmental Activism and World Civic Politics," *World Politics* 47 (April 1995): 311–340.
26. Virginia Hauffler, *A Public Role for the Private Sector* (Washington, D.C.: Carnegie Endowment for International Peace, 2001).
27. Adam Segal, *Digital Dragons* (Ithaca, NY: Cornell University Press, 2003).
28. Richard Doner, "Limits of State Strength: Toward an Institutionalist View of Economic Development," *World Politics* 44, no. 3 (1992): 398–431.
29. John Zysman, *Governments, Markets, and Growth* (Ithaca, NY: Cornell University Press, 1983).
30. Peter Evans, *Embedded Autonomy: States and Industrial Transformation* (Princeton, NJ: Princeton University Press, 1995).
31. Jon Pevehouse, "Democratization, Credible Commitments, and Joining International Organizations," in *Locating the Proper Authorities: The Interaction of Domestic and International Institutions,* ed. Daniel Drezner (Ann Arbor: University of Michigan Press, 2003), 25–48.
32. Sheila Jasanoff, *Designs on Nature: Science and Democracy in Europe and the United States* (Princeton, NJ: Princeton University Press, 2005).
33. Richard Rosecrance, *The Rise of the Virtual State* (New York: Basic Books, 1999).
34. Susan Strange, *The Retreat of the State: The Diffusion of Power in the World Economy* (New York: Cambridge University Press, 1996).
35. Thomas Friedman, *The Lexus and the Olive Tree,* new and expanded edition (New York: Anchor Books, 2000).
36. Daniel Yergin and Joseph Stanislaw, *The Commanding Heights: The Battle Between Government and the Marketplace that is Remaking the Modern World* (New York: Simon & Schuster, 1998).
37. Dani Rodrik, *Has Globalization Gone Too Far?* (Washington, D.C.: Institute for International Economics, 1997).
38. Alan Tonelson, *The Race to the Bottom* (Boulder, CO: Westview Press, 2000); and Jan Aart Scholte, *Globalization: A Critical Introduction* (New York: St. Martins Press, 2000).
39. Philip Cerny, "Globalization and the Erosion of Democracy," *European Journal of Political Research* 36, no. 1 (1999): 1–26.

40. Joseph Stiglitz, *Globalization and Its Discontents* (New York: W. W. Norton, 2003).
41. Norman Ellstrand, *Dangerous Liaisons? When Cultivated Plants Mate with Their Wild Relatives* (Baltimore: Johns Hopkins University Press, 2003).
42. Peter Andreas and Thomas Biersteker, eds., *The Rebordering of North America* (New York: Routledge, 2003); and Timothy Luke and Gearoid Tuanthail, "The Fraying Modern Map: Failed States and Contraband Capitalism," unpublished paper, undated.
43. Samuel Huntington, *The Clash of Civilizations and the Remaking of World Order* (New York: Simon & Schuster, 1996).

Chapter 2

1. Ian Wilmut, Keith Campbell, and Colin Tudge, *The Second Creation: The Age of Biological Control by the Scientists that Cloned Dolly* (London: Headline, 2000). Also see Anne McLaren, "The Decade of the Sheep," *Nature* 403, no. 6769 (2000): 479.
2. Declan Butler, "French Scientists Offered Time to Set Up Companies," *Nature* 397, no. 6716 (1999): 187.
3. Sheldon Krimsky and others, "Financial Interests of Authors in Scientific Journals," *Science and Engineering Ethics* 2, no. 4 (1996): 395–410.
4. Yonhap News Agency, "Korean Air Gives Cloning Expert 10-Year Free Flights," June 3, 2005, 1.
5. Dennis Normile and Charles Mann, "Asia Jockeys for Stem Cell Lead," *Science* 307, no. 5710 (2005): 662.
6. *Science*, "Biotechnology Start-Ups in Singapore: Inspiring Future Entrepreneurs," 295, no. 5563 (2002).
7. Emanuel Savas, *Privatization and Public-Private Partnerships* (New York: Chatham House, 2000); and Lawrence Rothenberg, *Regulation, Organizations, and Politics* (Ann Arbor: University of Michigan Press, 1994).
8. Donald Savoie, *Thatcher, Reagan, Mulroney: In Search of a New Bureaucracy* (Pittsburgh: University of Pittsburgh Press, 1994).
9. Haynes Johnson, *Sleepwalking Through History: America in the Reagan Years* (New York: Norton, 2003).
10. Darrell M. West, *Congress and Economic Policymaking* (Pittsburgh: University of Pittsburgh Press, 1987).
11. Martha Derthick and Paul Quirk, *The Politics of Deregulation* (Washington, D.C.: Brookings Institution, 1985).
12. David Stockman, *The Triumph of Politics: How the Reagan Revolution Failed* (New York: Harper & Row, 1986).
13. Kenneth Harris, *Thatcher* (Boston: Little Brown, 1988).
14. Peter Riddell, *The Thatcher Era and Its Legacy* (Oxford: Blackwell, 1991).
15. Organisation for Economic Co-Operation and Development, *Main Science and Technology Indicators, 2002* (Paris: OECD, 2002). Table 4–29 provides data on R & D changes from 1981 to 2000, while Table 4–26 shows the government and industry percent of R & D for the largest countries.

16. Melody Petersen, "A Conversation with Sheldon Krimsky: Uncoupling Campus and Company," *New York Times,* September 23, 2003, F2.
17. Association of University Technology Managers, *AUTM Licensing Survey: FY 2002* (Northbrook, IL: AUTM, 2003).
18. Jennifer Washburn, *University, Inc.: The Corporate Corruption of Higher Education* (New York: Basic Books, 2005), 49–53 and 168–69.
19. National Science Board, *Science and Engineering Indicators, 2004* (Arlington, VA: National Science Foundation, 2004), Appendix Table 5.2.
20. Sheldon Krimsky, *Science in the Private Interest* (Oxford: Rowman & Littlefield, 2003), 80–81.
21. Washburn, *University, Inc.: The Corporate Corruption,* 3–16.
22. Ibid., 140–41.
23. Ibid.
24. Albert Link, *A Generosity of Spirit: The Early History of the Research Triangle Park* (Research Triangle Park: Research Triangle Foundation of North Carolina, 1995).
25. David Gibson and Raymond Smilor, "The Role of the Research University in Creating and Sustaining the U.S. Technopolis," in *University Spin-off Companies: Economic Development, Faculty Entrepreneurs, and Technology Transfer,* eds. Alistair Brett, David Gibson, and Raymond Smilor (Savage, MD: Rowman & Littlefield, 1991), 31–70.
26. Luc Anselin, Attila Varga, and Zoltan Acs, "Local Geographic Spillovers Between University Research and High Technology Innovations," *Journal of Urban Economics* 42, no. 3 (1997): 422–448; and Michael Luger and Harvey Goldstein, *Technology in the Garden: Research Parks and Regional Economic Development* (Chapel Hill: University of North Carolina Press, 1991).
27. Alistair Brett, David Gibson and Raymond Smilor, eds., *University Spin-off Companies: Economic Development, Faculty Entrepreneurs, and Technology Transfer* (Savage, MD: Rowman & Littlefield, 1991), xxii.
28. Sheila Jasanoff, *The Fifth Branch: Science Advisers as Policymakers* (Cambridge, MA: Harvard University Press, 1990).
29. Sheila Jasanoff, *Designs on Nature: Science and Democracy in Europe and the United States* (Princeton: Princeton University Press, 2005), 149–50.
30. Jasanoff, *The Fifth Branch: Science Advisers.*
31. Sandra Laville and Duncan Campbell, "Animal Rights Extremists in Arson Spree," *Guardian,* July 1, 2005, 10.
32. Jamie Shreeve, "The Other Stem-Cell Debate," *New York Times Magazine,* April 10, 2005, 44.
33. Normile and Mann, "Asia Jockeys," *Science.*
34. Dennis Normile, "Can Money Turn Singapore into a Biotech Juggernaut?" *Science* 297, no. 5586 (2002).
35. Ibid.
36. Martha Overland, "A Tale of 2 Countries: Singapore's Regeneration," *Chronicle of Higher Education,* November 11, 2005, A42–46.
37. National Science Board, *Science and Engineering Indicators 2004* (Washington, D. C.: National Science Foundation, 2004), 0–6.

38. David Barboza, "China's Problem with 'Anti-Pest' Rice," *New York Times,* April 16, 2005, B1.
39. Namrata Nadkarni, "US Exports of GM Modified Corn Hit a New Setback: Japan May Ship Back or Destroy Unauthorised Cargo," *Lloyd's List,* June 28, 2005, 4.

Chapter 3

1. *Merriam-Webster's Collegiate Dictionary, http://search.eb.com/dictionary.*
2. Centers for Disease Control and Prevention, "Assisted Reproductive Technology," www.cdc.gov/ART/index.htm.
3. Ibid.
4. Debora Spar, *The Baby Business: How Money, Science, and Politics Drive the Commerce of Conception* (Cambridge, MA: Harvard Business School Press, 2006).
5. Jennifer Strickler, "The New Reproductive Technology," *Sociology of Health & Illness* 14, no. 1 (1992): 111–32.
6. David Dunson, Donna Baird, and Bernardo Colombo, "Increased Infertility With Age in Men and Women," *Obstetrics & Gynecology* 103 (2004): 51–56.
7. InterNational Council on Infertility Information Dissemination, "A History of IVF Statistics," www.inciid.org.
8. Meredith Reynolds and Laura Schieve, "Insurance Coverage and Outcomes of In Vitro Fertilization," *New England Journal of Medicine* 348, no. 10 (2003): 958–59.
9. Vishvanath Karande and others, "Prospective Randomized Trail Comparing the Outcome and Cost of In Vitro Fertilization with that of a Traditional Treatment Algorithm as First-Line Therapy for Couples with Infertility," *Fertility and Sterility* 71, no. 3 (1999): 468–75.
10. 20th Century History, "First Test-Tube Baby—Louise Brown," www.history 1900s.about.com.
11. Carol Warner, "Research Description," www.biology.neu.edu/faculty03/ warner03.html.
12. Marie-Victoire Senat and others, "How Does Multiple Pregnancy Affect Maternal Mortality and Morbidity?" *Clinical Obstetrics and Gynecology* 41, no. 1 (1998): 79–83.
13. Lyria Bennett Moses, "Understanding Legal Responses to Technological Change: The Example of In Vitro Fertilization," *Minnesota Intellectual Property Review,* May 2005.
14. Cynthia Cohen, "Unmanaged Care: The Need to Regulate New Reproductive Technologies in the United States," *Bioethics* 11, no. 3 & 4 (1997): 348–65.
15. Ibid.
16. Heather Mason Kiefer, "The Birth of In Vitro Fertilization," *The Gallup Poll,* August 5, 2003.
17. Jennifer Robison, "Infertile Women Quest for Family," *The Gallup Poll,* May 21, 2002.

18. Le Moyne College/Zogby International, "Contemporary Catholic Trends," November 16, 2001. www.Zogby.com.
19. Rick Weiss, "Babies in Limbo: Laws Outpaced by Fertility Advances," *Washington Post,* February 8, 1998, A1.
20. Mike Allen and Rick Weiss, "Bush Rejects Stem Cell Compromise," *Washington Post,* May 26, 2005, A2.
21. Joseph Schenker, "Assisted Reproduction Practice in Europe: Legal and Ethical Aspects," *Human Reproduction Update* 3, no. 2 (1997): 173–84.
22. Sheila Jasanoff, *The Fifth Branch: Science Advisers as Policymakers* (Cambridge, MA: Harvard University Press, 1990).
23. Schenker, "Assisted Reproduction in Europe."
24. Ivar Bleiklie, "Governing Assisted Reproductive Technology: The Case of Norway" (paper prepared for delivery at the annual meeting of the American Political Science Association, San Francisco, August 30–September 2, 2001).
25. Alison Abbott, "Germany's Past Still Casts a Long Shadow," *Nature* 389, no. 6652 (1997): 660.
26. Darren Langdridge and Eric Blyth, "Regulation of Assisted Conception Services in Europe," *Journal of Social Welfare and Family Law* 23, no. 1 (2001): 45–64.
27. *Associated Press,* "Swiss Voters Reject New Initiatives," December 3, 2000.
28. Ulrich Bahnsen, "Swiss to Vote on Ban on In Vitro Fertilization," *Nature* 396, no. 6707 (1998): 105.
29. Martin Johnson, "Should the Use of Assisted Reproduction Techniques be Deregulated?" *Human Reproduction* 13, no. 7 (1998): 1769–76.
30. Sarah Boseley, "Fertility Debate: Public Asked to Help Rewrite IVF Law," *The Guardian,* August 17, 2005, 4.
31. A. Nyboe Andersen and others, "Assisted Reproductive Technology in Europe, 2001," *Human Reproduction* 20, no. 5 (2005): 1158–76.
32. Elizabeth Heitman, "Social and Ethical Aspects of In Vitro Fertilization," *International Journal of Technology Assessment in Health Care* 15, no. 1 (1999): 22–35.
33. Ibid.
34. Dan Bilefsky, "Court Rules Couple Must Agree on Use of Embryos," *International Herald-Tribune,* March 8, 2006, 1.
35. Michael Barnhart, "Nature, Nurture, and No-Self: Bioengineering and Buddhist Values," *Journal of Buddhist Ethics* 7 (2000).
36. Pankaj Mishra, "How India Reconciles Hindu Values and Biotech," *New York Times,* August 21, 2005, 4.
37. Ibid.
38. Darryl Macer, J. Azariah, and P. Srinives, "Attitudes to Biotechnology in Asia," *International Journal of Biotechnology* 2 (2000): 313–32.
39. Gabor Kovacs and others, "Community Attitudes to Assisted Reproductive Technology: a 20-Year Trend," *Medical Journal of Australia* 179, no. 10 (2003): 536–38.

40. Darryl Macer, "Perception of Risks and Benefits of In Vitro Fertilization, Genetic Engineering and Biotechnology," *Social Science and Medicine* 38 (1994): 23–33.
41. Ibid.
42. Ren-Zong Qiu, "Sociocultural Dimensions of Infertility and Assisted Reproduction in the Far East," online at www.who.int/reproductive-health/infertility/12.pdf, undated.
43. Ole Doring, "China's Struggle for Practical Regulations in Medical Ethics," *Nature* 4 (March 2003): 233–39.
44. Ya Zhang, "The Mutual Interaction Between Genetic Information Technology and Societal Factors in China" (unpublished paper, School of IST, Penn State University, 2003).
45. Ernest Hung Yu Ng and others, "Regulating Reproductive Technology in Hong Kong," *Journal of Assisted Reproduction and Genetics* 20, no. 7 (2003): 281–86.
46. Aditya Bharadwaj, "How Some Indian Baby Makers Are Made: Media Narratives and Assisted Conception in India," *Anthropology and Medicine* 7, no. 1 (2000): 63–78.
47. Ganapati Mudur, "India Considers Government Agency to License Infertility Clinics," *British Medical Journal* 325 (September 14, 2002): 564.
48. Anjali Widge, "Sociocultural Attitudes Towards Infertility and Assisted Reproduction in India," online at www.who.int/reproductive-health/infertility/11.pdf, undated.
49. Celina Ramjoue, "Assisted Reproductive Technology Policy in Italy" (paper presented at the ECPR Joint Sessions Workshop, Turin, Italy, March 22–27, 2002).
50. Congregation for the Doctrine of the Faith, "Instruction on Respect for Human Life in its Origin and on the Dignity of Procreation Replies to Certain Questions of the Day," online at www.vatican.va, February 22, 1987.
51. Ramjoue, "Assisted Reproductive Technology."
52. Ibid.
53. Andrea Boggio, "Italy Enacts New Law on Medically Assisted Reproduction," *Human Reproduction* 20, no. 5 (2005): 1153–57.
54. Elisabeth Rosenthal and Elisabetta Povoledo, "Vote on Fertility Law Fires Passions in Italy," *New York Times,* June 11, 2005, 7.
55. Sophie Arie, "In Europe, Italy Now a Guardian of Embryo Rights," *Christian Science Monitor,* June 14, 2005, 1.
56. Susan Mitchell, "We're having an IVF Baby," *Financial Times,* September 19, 2004.
57. *Irish Times,* "Public Divided on Infertility Treatment Ethics," May 13, 2005, 9.
58. Ibid.; and Liam Reid and Carol Coulter, "Regulatory Body for Fertility Treatment Urged," *Irish Times,* March 13, 2006, 8.
59. Mitchell, "We're having an IVF Baby."
60. Eithne Donnellan, "Couples Going Abroad for Tests on Embryos," *Irish Times,* September 16, 2005, 3.
61. *Irish Times,* "Public Divided."

62. John Robertson, "Assisted Reproduction in Germany and the United States," *Berkeley Electronic Press,* Paper 226, April 1, 2004, 1–46.

Chapter 4

1. *Wall Street Journal,* "Biotech Crops Grew at Slowest Pace Since 1996," January 12, 2006, A11; and *New York Times,* "Group Reports Increase in Biotech Harvest," January 13, 2005, C2.
2. David Vogel and Diahanna Lynch, "The Regulation of GMOs in Europe and the United States" (paper presented to the Council on Foreign Relations, New York, April 5, 2001).
3. Akihiro Hino, "Safety Assessment and Public Concerns for Genetically Modified Food Products: The Japanese Experience," *Toxicologic Pathology* 30, no. 1 (2002): 126–28.
4. U.S. Food and Drug Administration, "FDA's Policy for Foods Developed by Biotechnology" (unpublished 1995 report online at www.cfsan.fda.gov).
5. Nina Fedoroff and Nancy Brown, *Mendel in the Kitchen: A Scientist's View of Genetically Modified Foods* (Washington, D.C.: Joseph Henry Press, 2004).
6. Hino, "Safety Assessment and Public Concerns."
7. Thomas Bernauer and Erika Meins, "Scientific Revolution Meets Policy and the Market: Explaining Cross-National Differences in Agricultural Biotechnology Regulation" (Discussion Paper, Centre for International Economic Studies 1445–3746, No. 144, Adelaide, November 2001).
8. Ibid.
9. Murray Fulton and Konstantinos Giannakas, "Agricultural Biotechnology and Industry Structure," *AgBioForum* 4, no. 2 (2001): 137–51.
10. Bill Lambrecht and Deirdre Shesgreen, "Monsanto Lobbies to Keep the Status Quo for Gene-Altered Crops," *St. Louis Post-Dispatch,* September 11, 2005, A9.
11. Rob Horsch and Jill Montgomery, "Why We Partner: Collaborations Between the Private and Public Sectors for Food Security and Poverty Alleviation through Agricultural Biotechnology," *AgBioForum* 7, no. 1 & 2 (2004): 80–83.
12. Gina Kolata, "Pork That's Good for the Heart May Be Possible with Cloning," *New York Times,* March 27, 2006, A1.
13. Bernauer and Meins, "Scientific Revolution Meets Policy."
14. U.S. Food and Drug Administration, "Safety Assurance of Foods Derived by Modern Biotechnology in the United States" (unpublished July 1996 presentation online at www.cfsan.fda.gov).
15. Ibid.
16. U.S. Food and Drug Administration, "FDA's Policy for Foods."
17. Ibid.
18. Ibid.
19. Andrew Pollack, "Lax Oversight Found in Tests of Gene-Altered Crops," *New York Times,* January 3, 2006, D2.
20. Fedoroff and Brown, *Mendel in the Kitchen;* and Michael Taylor and Jody Tick, "The StarLink Case: Issues for the Future" (unpublished paper prepared by

Resources for the Future and the Pew Initiative on Food and Biotechnology, undated).

21. Pew Initiative on Food and Biotechnology, "Recent Poll Findings," August 5–10, 2003 survey online at http://pewagbiotech.org.
22. William Hallman and others, *Public Perceptions of Genetically Modified Foods* (New Brunswick, NJ: Food Policy Institute, October 2003).
23. Genetically Engineered Organisms Public Issues Education Project, *GE Foods in the Market* (Ithaca, NY: Cornell Cooperative Extension, 2003).
24. Hallman and others, *Public Perceptions.*
25. Pew Initiative "Recent Poll Findings."
26. *Wall Street Journal,* "Biotech Crops Grew."
27. Kristin Rosendal, "Governing GMOs in the EU: A Deviant Case of Environmental Policy-Making?" *Global Environmental Politics,* Vol. 5, no. 1 (2005): 82.
28. Andrew Pollack, "Trade Ruling is Expected to Favor Biotech Food," *New York Times,* February 6, 2006, C6.
29. Edward Alden and Jeremy Grant, "WTO Rules Against Europe in GM Food Case," *Financial Times,* February 8, 2006, 6.
30. Tom Wright, "Swiss Voters Approve Ban on Genetically Altered Crops," *New York Times,* November 28, 2005, C2.
31. Scott Miller and Joel Clark, "EU Orders Greece to Lift Gene-Altered Seed Ban," *Wall Street Journal,* January 11, 2006, A13.
32. Paul Meller, "Europe Rejects Looser Labels for Genetically Altered Food," *New York Times,* September 9, 2004, 7.
33. Allison Abbott and Burkhardt Roeper, "Germany Seeks 'Non-Modified' Food Label," *Nature* 391, no. 6670 (1998): 828.
34. Meller, "Europe Rejects Looser Labels."
35. *Europe Agri,* "Genetic Engineering: Madrid Plans Safety Distances Between Traditional and GM Crops," December 21, 2004, 1.
36. Ehsan Masood, "Royal Society Wants Genetics Watchdog," *Nature* 395, no. 6697 (1998): 5.
37. *Turkish Daily News,* "Turkey Working to Form Policy on Genetically Modified Organisms," May 25, 2005, 1.
38. Paul Meller, "Europeans to Toughen Rules on Animal Feed From U.S.," *New York Times,* April 13, 2005, C4.
39. Bernauer and Meins, "Scientific Revolution Meets Policy."
40. George Gaskell and others, "Worlds Apart? The Reception of Genetically Modified Foods in Europe and the U.S.," *Science* 285, no. 5426 (1999): 384–87.
41. Ibid.
42. Nigel Williams, "Agricultural Biotech Faces Backlash in Europe," *Science* 281, no. 5378 (1998): 768–71.
43. Michael Burton and others, "Consumer Attitudes to Genetically Modified Organisms in Food in the UK," *European Review of Agricultural Economics* 28, no. 4 (2001): 479–98.
44. Karma Hickman, "Biotech Industry Calls for GMO Education," *ANSA English Media Service,* March 14, 2005.

45. Alex Scott, "Brussels Presses EU Members to Hasten GM Approvals," *Chemical Week,* March 30, 2005, 17.
46. Thomas McDevitt, *World Population Profile: 1998* (Washington, D.C.: U.S. Government Printing Office, 1999), A12.
47. Randy Hautea and Margarita Escaler, "Plant Biotechnology in Asia," *AgBioForum* 7, no. 1 & 2 (2004): Article 1.
48. Ibid.
49. Ibid.
50. Roderick dela Cruz, "UNCTAD Calls on RP, Other Countries to Balance Impact of GMO," *Manila Standard,* May 17, 2005.
51. Xiong Lei, "China Could be First Nation to Approve Sale of GM Rice," *Science* 306, no. 5701 (2004): 1458–59.
52. *Business Daily Update,* "GM Rice May Soon be Commercialized," January 28, 2005, section 28.
53. Geoff Dyer, "Controversy Grows Over China's Biotech Crops," *Financial Times,* June 24, 2005, 14.
54. Jikun Huang, Qinfang Wang, and James Keeley, "Agricultural Biotechnology Policy Processes in China" (unpublished paper, Institute for Development Studies, University of Sussex, 2001), at http://www.ids.ac.uk/idS/KNOTS/Projects/biotech/pubsPolproc.html.
55. Ibid.
56. Mary Marchant, Cheng Fang, and Baohui Song, "Issues on Adoption, Import Regulations, and Policies for Biotech Commodities in China with a Focus on Soybeans," *AgBioForum* 5, no. 4 (2002): 167–74.
57. *Business Daily* "GM Rice May Soon."
58. *BBC Monitoring International Reports,* "China Ratifies GMO Trade Protocol," May 19, 2005.
59. Sabrina Safrin, "Treaties in Collision? The Biosafety Protocol and the World Trade Organization Agreements," *American Journal of International Law* 96 (2002): 606–28.
60. Hautea and Escaler, "Plant Biotechnology in Asia."
61. *Agence France Presse,* "China Planning Large-Scale Introduction of Genetically Engineered Rice," November 30, 2004.
62. David Barboza, "China's Problem With 'Anti-Pest' Rice," *New York Times,* April 16, 2005, B1.
63. David Barboza, "Modified Rice May Benefit China Farms, Study Shows," *New York Times,* May 3, 2005, C8.
64. Barboza, "China's Problem With 'Anti-Pest' Rice."
65. *Agence France Presse,* "China Planning."
66. James Randerson, "By the People, For the People: Genetically Modified Crops Don't Just Come from Big Corporations," *New Scientist* (February 19, 2005): 36.
67. Magdy Madkour, Latha Nagarajan, and Clemen Gehlhar, "Private-Public Partnerships in Bringing BioTechnology from Laboratory to the Market Place: Comparison of Egypt and India" (paper presented at 6th International ICABR Conference, Ravello, Italy, July 11–14, 2002).

68. Hautea and Escaler, “Plant Biotechnology in Asia.”
69. Madkour, Nagarajan, and Gehlhar, “Private-Public Partnerships.”
70. John Feffer, “Asia Holds Key to Future of GM Food,” *Korea Herald,* December 6, 2004.
71. K. S. Jayaraman, “India Approves Use of Genetically Modified Crops, Despite Critics,” *Nature* 397, no. 6716 (1999): 188.
72. Colin Carter and Guillaume Gruere, “International Approaches to the Labeling of Genetically Modified Foods,” *Choices* (Second Quarter, 2003): 1–4.
73. Hino, “Safety Assessment and Public Concerns.”
74. Feffer, “Asia Holds Key.”
75. Namrata Nadkarni, “US Exports of GM Modified Corn Hit a New Setback,” *Lloyd’s List,* June 28, 2005, 4.
76. Kazuo Watanebe, Mohammad Taeb, and Haruko Okusu, “Japanese Controversies over Transgenic Crop Regulation,” *Science* 305, no. 5690 (2004): 1572.
77. Masakazu Inaba and Darryl Macer, “Policy, Regulation and Attitudes towards Agricultural Biotechnology in Japan,” *Journal of International Biotechnology Law* 1 (2004).
78. Darryl Macer, J. Azariah, and P. Srinives, “Attitudes to Biotechnology in Asia,” *International Journal of Biotechnology* 2 (2000): 313–32.
79. Peter Phillips and Heather McNeill, “A Survey of National Labeling Policies for GM Foods,” *AgBioForum* 3, no. 4 (2000): article 7.
80. Erik Millstone, Eric Brunner, and Sue Mayer, “Beyond ‘Substantial Equivalence,’” *Nature* 401, no. 6753 (1999).
81. Phillips and McNeill, “A Survey of National Labeling Policies.”
82. Ian Sheldon, “Regulation of Biotechnology: Will We Ever ‘Freely’ Trade GMOs?” *European Review of Agricultural Economics* 29, no. 1 (2002): 155–76.

Chapter 5

1. Arlene Klotzko, *A Clone of Your Own? The Science and Ethics of Cloning* (New York, NY: Oxford University Press, 2004).
2. Gayle Woloschak, “Transplantation: Biomedical and Ethical Concerns Raised by the Cloning and Stem-Cell Debate,” *Zygon* 38, no. 3 (2003): 699–704.
3. Ruben Lisker, “Ethical and Legal Issues in Therapeutic Cloning and the Study of Stem Cells,” *Archives of Medical Research* 34, no. 6, (2003): 607–11.
4. Steven Stice, “Animal Cloning, Enhancements, and Genetic Selection” (paper presented at the 38th Annual Veterinary Conference, University of Georgia, April 6–8, 2001), at http://www.georgiacenter.uga.edu/conferences/2001/Apr/06/vet_alumni.phtml.
5. Donald Bruce, “Polly, Dolly, Megan, and Morag: A View from Edinburgh on Cloning and Genetic Engineering,” *Philosophy and Technology* 3, no. 2 (1997): 37–52.
6. H. J. Webber, “New Horticultural and Agricultural Terms,” *Science* 28 (1903): 501–3.

7. Leyla Dinc, "Ethical Issues Regarding Human Cloning," *Nursing Ethics* 10, no. 3 (2003): 238–54.
8. L. R. Sanchez-Sweatman, "Reproductive Cloning and Human Health: An Ethical, International, and Nursing Perspective," *International Nursing Review* 47, no. 1, (2000): 28.
9. Richard Monastersky, "A Second Life for Cloning," *Chronicle of Higher Education,* February 3, 2006, A14–17.
10. Dinc, "Ethical Issues."
11. Shaun Pattinson and Timothy Caulfield, "Variations and Voids: The Regulation of Human Cloning Around the World," *BMC Medical Ethics* 5, no. 9 (2004).
12. Leon Kass, "The Wisdom of Repugnance: Why We Should Ban the Cloning of Humans," *New Republic* 216, no. 22 (1997).
13. John Evans, "Cloning Adam's Rib: A Primer on Religious Responses to Cloning" (report prepared for the Pew Forum on Religion and Public Life, undated).
14. Steven Best and Douglas Kellner, "Biotechnology, Ethics and the Politics of Cloning," *Democracy and Nature* 8, no. 3 (2002): 439–65.
15. Ian Wilmut, Keith Campbell, and Colin Tudge, *The Second Creation: Dolly and the Age of Biological Control* (New York, NY: Farrar, Straus and Giroux, 2000).
16. Rudolf Jaenisch and Ian Wilmut, "Don't Clone Humans!" *Science* 291, no. 5513 (2001): 2552.
17. Pattinson and Caulfield, "Variations and Voids."
18. Colum Lynch, "U.N. Backs Human Cloning Ban," *Washington Post,* March 9, 2005, A15.
19. *Nature,* "UN Compromise Ends Human Cloning Debate with 'Non-Binding' Ban" 434, no. 7031 (2005): 264.
20. Sanchez-Sweatman, "Reproductive Cloning and Human Health."
21. Cynthia Fox, "Cloning Laws, Policies, and Attitudes Worldwide," *IEEE Engineering in Medicine and Biology* (March/April, 2004): 55–61.
22. Jane Maienschein, "What's in a Name: Embryos, Clones, and Stem Cells," *American Journal of Bioethics* 2 (2002): 12–19.
23. Carla Messikomer, Renee Fox, and Judith Swazey, "The Presence and Influence of Religion in American Bioethics," *Perspectives in Biology and Medicine* 44, no. 4 (2001): 485–508.
24. Louise Bernier and D. Gregoire, "Reproductive and Therapeutic Cloning, Germline Therapy, and Purchase of Gametes and Embryos: Comments on Canadian Legislation Governing Reproduction Technologies," *Journal of Medical Ethics* 30 (2004): 527–32.
25. President's Council on Bioethics, *Human Cloning and Human Dignity: An Ethical Inquiry* (Washington, D.C., 2002); and Mary Mahowald, "The President's Council on Bioethics, 2002–2004," *Perspectives on Biology and Medicine* 48 (2005): 159–71.
26. Helen Pearson, "Cloning Success Marks Asian Nations as Scientific Tigers," *Nature* 427, no. 6976 (2004): 664.
27. Matthew Nisbet, "Public Opinion about Stem Cell Research and Human Cloning," *Public Opinion Quarterly* 68, no. 1 (2004): 131–54.

28. Ibid.
29. Frederic Frommer, "Dairy Industry, Consumers Skeptical of Cloned Cows," *Providence Journal,* July 13, 2005, E8.
30. Ibid.
31. Cesare Galli and others, "A European Perspective on Animal Cloning and Government Regulation," *IEEE Engineering in Medicine and Biology* (March/April, 2004): 52–54.
32. Ibid.
33. Best and Kellner, "Biotechnology, Ethics and Politics of Cloning."
34. Fox, "Cloning Laws, Policies, and Attitudes Worldwide."
35. Sarah Parry, "The Politics of Cloning: Mapping the Rhetorical Convergence of Embryos and Stem Cells in Parliamentary Debates," *New Genetics and Society* 22, no. 2 (2003): 145–68.
36. Roger Highfield, "Embryo Cloning for Diabetes Research," *London Daily Telegraph,* April 20, 2005, 12.
37. Sarah-Kate Templeton, "Dying Briton Pins Last Hope on Cloning," *London Sunday Times,* May 22, 2005, 12.
38. Fox, "Cloning Laws, Policies, and Attitudes Worldwide."
39. Brad Spurgeon, "France Bans Reproductive and Therapeutic Cloning," *British Medical Journal* 329 (July 17, 2004).
40. Galli and others, "A European Perspective."
41. Brian Salter and Mavis Jones, "Regulating Human Genetics: The Changing Politics of Biotechnology Governance in the European Union," *Health, Risk & Society* 4, no. 3 (2002): 325–40.
42. "Public Opinion: Detailed Survey Results," January 2003 EOS Gallup Europe survey online at www.genetics-and-society.org/analysis/opinion/detailed.html.
43. Ibid.
44. Darryl Macer, "Bioethics in Asia," in *Encyclopedia of the Human Genome* (London: Nature MacMillan, 2003), 277–80.
45. Gerhold Becker, "Cloning Humans? The Chinese Debate and Why It Matters," *Eubios Journal of Asian and International Bioethics* 7 (1997): 175–78.
46. *Nature,* "China's Human-Cloning Policy Fudges Law on Cross-Species Fusions," 427, no. 6972 (2004): 278.
47. Michael Frith, "Asian Nations Approach Cloning Consensus," *Nature Medicine* 248 (2003).
48. Margaret Munro, "S. Koreans Tailor Stem Cells to Fit Disease," *Montreal Gazette,* May 20, 2005, A25.
49. Yoo Soh-jung, "Government to Establish Global Consortium for Cloning Research," *Korea Herald,* May 23, 2005.
50. Indian Council of Medical Research, *Ethical Guidelines for Biomedical Research on Human Subjects* (New Delhi, India: ICMR, 2000), 48.
51. Frith, "Asian Nations Approach Cloning Consensus."
52. Ibid.
53. Fox, "Cloning Laws, Policies, and Attitudes Worldwide."
54. Asako Saegusa, "Japan's Bioethics Debate Lags Behind Thinking in the West," *Nature* 389, no. 6652 (1997): 661.

55. Patrick Hopkins, "Bad Copies: How Popular Media Represent Cloning as an Ethical Problem," *Hastings Center Report* 28, no. 2 (1998).
56. Elisabeth Rosen, "The Dolly-Dollar Dichotomy: Animal Cloning Restrictions and the Competitiveness of the European Biotech Industry," *Nordic Journal of International Law* 67 (1998): 423–30.

Chapter 6

1. *Asia Africa Intelligence Wire*, "Task Force to Seek Funds for Stem Cell Research," April 6, 2005.
2. Matthew Nisbet, "The Controversy Over Stem Cell Research and Medical Cloning," Committee for the Scientific Investigation of Claims of the Paranormal, unpublished paper, April 2, 2004.
3. Anne McLaren, "Ethical and Social Considerations of Stem Cell Research," *Nature* 414, no. 6859 (2001): 129–31.
4. Denise Stevens, "Embryonic Stem Cell Research," *Houston Journal of International Law* (Spring 2003).
5. Eric Juengst and Michael Fossel, "The Ethics of Embryonic Stem Cells—Now and Forever, Cells Without End," *JAMA* 284, no. 24 (2000): 3180–84.
6. National Research Council and Institute of Medicine, *Guidelines for Human Embryonic Stem Cell Research* (Washington, D.C.: National Academies Press, 2005).
7. Stevens, "Embryonic Stem Cell Research."
8. Tony Reichardt, "Studies of Faith," *Nature* 432, no. 7018 (2004): 666–69.
9. Dietmar Mieth, "Going to the Roots of the Stem Cell Debate: The Ethical Problems of Using Embryos for Research," *EMBO Reports* 1, no. 1 (2000): 4–6.
10. Leon Kass, "The Wisdom of Repugnance," *New Republic* 216, no. 22 (1997).
11. Claudia Dreifus, "At Harvard's Stem Cell Center, the Barriers Run Deep and Wide," *New York Times*, January 24, 2006, D2.
12. Alexander Capron, "Stem Cells: Ethics, Law and Politics," *Biotechnology Law Report* 20, no. 5 (2001): 678–97.
13. President's Council on Bioethics, *Monitoring Stem Cell Research* (Washington, D.C.: President's Council on Bioethics, 2004).
14. President George Bush, "I Have Given this Issue a Great Deal of Thought," *Science* 293, no. 5533 (2001): 1245.
15. Dreifus, "At Harvard's Stem Cell Center."
16. Sheryl Stolberg, "House Approves a Stem Cell Bill Fought by Bush," *New York Times*, May 2, 2005, A1.
17. *National Review*, "From the House, a Disgrace," June 20, 2005.
18. Elisabeth Rosenthal, "Britain Embraces Embryonic Stem Cell Research," *New York Times*, August 24, 2004, 6.
19. National Research Council and Institute of Medicine, *Guidelines for Human Embryonic Stem Cell Research* (Washington, D.C.: National Academies Press, 2005).
20. Peter Aldhous, "After the Gold Rush," *Nature* 434 no. 7034 (2005): 694–96.

21. Andrew Pollack, "A Case of Stunted Growth: California's Stem Cell Program is Hobbled but Staying the Course," *New York Times,* December 10, 2005, B1.
22. Matthew Nisbet, "Public Opinion About Stem Cell Research and Human Cloning, *Public Opinion Quarterly* 68, no. 1 (2004): 131–54.
23. Ibid.
24. Lydia Saad, "Most Americans in the Dark About Stem Cell Research," *Gallup Poll,* July 20, 2001.
25. Julie Ray, "Medicine Meets Morality in Stem Cell Debate," *Gallup Poll,* August 5, 2003.
26. Line Matthiessen-Guyader, "Survey on Opinions from National Ethics Committees or Similar Bodies, Public Debate and National Legislation in Relation to Human Embryonic Stem Cell Research and Use" (European Commission report, July 2004).
27. Allison Ayer, "Stem Cell Research: The Laws of Nations and a Proposal for International Guidelines," *Connecticut Journal of International Law* (Spring 2002).
28. Richard Stone, "U.K. Backs Use of Embryos, Sets Vote," *Science* 289, no. 5483, (2000): 1269–70; and Christine Hauskeller, "How Traditions of Ethical Reasoning and Institutional Processes Shape Stem Cell Research in Britain," *Journal of Medicine and Philosophy* 29, no. 5 (2004): 509–32.
29. Rosenthal, "Britain Embraces Embryonic Stem Cell Research."
30. Tim Webb, "Laboratory to the World: The UK's Big Push on Stem Cells," *London Independent,* March 20, 2005, 15.
31. Helen Lawton Smith, "Regulating Science and Technology: The Case of the UK Biotechnology Industry," *Law & Policy* 27, no. 1 (2005): 189–212.
32. Jan Beckmann, "On the German Debate on Human Embryonic Stem Cell Research," *Journal of Medicine and Philosophy* 29, no. 5 (2004): 603–21.
33. Angela Campbell, "Ethos and Economics: Examining the Rationale Underlying Stem Cell and Cloning Research Policies in the United States, Germany, and Japan," *American Journal of Law & Medicine* (2005).
34. Bertrand Benoit, "Schroeder in Call on Stem Cell Research," *Financial Times,* June 15, 2005, 9.
35. *United Press International* (UPI), "German Chancellor Backs Stem Cell Research," June 14, 2005.
36. Giovanni Maio, "The Embryo in Relationships: A French Debate on Stem Cell Research," *Journal of Medicine and Philosophy* 29, no. 5 (2004): 583–602.
37. Federica Castellani, "Europe's Stem Cell Workers Pull Together," *Nature* 432, no. 7015 (2004): 260.
38. Maio, "The Embryo in Relationships."
39. Franz Pichler, "Ethical Questions on the Funding of Human Embryonic Stem Cell Research in the Sixth Framework Programme of the European Commission for Research, Technological Development and Demonstration," *Innovation* 18, no. 2 (2005): 261–71.
40. Dennis Normile and Charles Mann, "Asia Jockeys for Stem Cell Lead," *Science* 307, no. 5710 (2005): 660–64.
41. Darryl Macer, "Asian Approaches to Stem Cell Research and IP Protection" (paper presented at the International Conference on "Bioethical Issues of

Intellectual Property in Biotechnology," Tokyo, Japan, 6–7 September 2004), at http://www.ipgenethics.org/conference.htm.

42. Dennis Normile and Charles Mann, "Asian Countries Permit Research, With Safeguards," *Science,* Vol. 307, no. 5710 (2005): 664.
43. LeRoy Walters, "Human Embryonic Stem Cell Research," *Kennedy Institute of Ethics Journal* 14, no. 1 (2004): 3–38.
44. *New Scientist,* "Stem Cell Standoff," June 4, 2005, 6.
45. Normile and Mann, "Asia Jockeys for Stem Cell Lead."
46. Gina Kolata, "Koreans Report Ease in Cloning for Stem Cells," *New York Times,* May 20, 2005, A1.
47. Nicholas Wade, "Korean Researchers to Help Others Clone Cells for Study," *New York Times,* October 19, 2005, A14.
48. *New York Times,* "Faked Research on Stem Cells is Confirmed by Korean Panel," December 23, 2005, A8; and Nicholas Wade, "Korean Scientist Said to Admit Fabrication in a Cloning Study," *New York Times,* December 16, 2005, A1; and Nicholas Wade and Choe Sang-Hun, "Human Cloning Was All Faked, Koreans Report," *New York Times,* January 10, 2006, A1.
49. Normile and Mann, "Asia Jockeys for Stem Cell Lead."
50. Ibid.
51. Ibid.
52. Robert MacGregor, "The Effects of Culture and Policy on Science in China" (unpublished paper, Rice University, May 4, 2004); and Bruce Einhorn, "A Cancer Treatment You Can't Get Here," *Business Week,* March 6, 2006, 88–90.
53. Yanguang Wang, "Chinese Ethical Views on Embryo Stem (ES) Cell Research," in *Bioethics in Asia in the 21st Century,* eds., S. Y. Song, Y. M. Koo, and D. Macer, (Tsukuba, Japan: Eubios Ethics Institute, 2003), 49–55.
54. Normile and Mann, "Asia Jockeys for Stem Cell Lead."
55. Stephen Seawright, "A New Report Says Mainland Stem Cell Research is Eclipsing the International Field, While Western Researchers Remain Bogged Down by Moral Concerns," *South China Morning Post,* February 24, 2005, 16.
56. Ibid.
57. Ole Doring, "Chinese Researchers Promote Biomedical Regulations," *Kennedy Institute of Ethics Journal* 14, no. 1 (2004): 39–46.
58. Seawright, "Mainland Stem Cell Research is Eclipsing."
59. Einhorn, "A Cancer Treatment You Can't Get Here."
60. *Hindustan Times,* "No Law, No Check on Stem Cell Clinics," March 18, 2005.
61. K. S. Jayaraman, "Indian Regulations Fail to Monitor Growing Stem Cell Use in Clinics," *Nature* 434, no. 7031 (2005): 259.
62. *India Business Insight,* "India Plans Fund for Stem Cell Research," April 7, 2005.
63. Sara Harris, "Asian Pragmatism: Japan Has Set up a Legal Framework to Allow the Use and Generation of Human Embryonic Stem Cell Lines," *EMBO Reports* 3, no. 9 (2002): 816–17.
64. Akira Akabayashi and Brian Slingsby, "Biomedical Ethics in Japan," *Cambridge Quarterly of Healthcare Ethics* 12 (2003): 261–64; and *Tokyo Daily Yomiuri,* "Japan Must Join Pursuit of Scientific Progress," May 23, 2005, 4.

65. Katy Human, "U.S. Researchers Recruited Labs in China, South Korea and England Are Attracting Scientists Frustrated by the Lack of Stem Cell Funding," *Denver Post,* May 25, 2005, A21.
66. Angela Campbell, "Ethos and Economics: Examining the Rationale Underlying Stem Cell and Cloning Research Policies in the United States, Germany, and Japan," *American Journal of Law & Medicine* (2005).
67. David Resnik, "Privatized Biomedical Research, Public Fears, and the Hazards of Government Regulation," *Health Care Analysis* 7 (1999): 273–87.
68. Ibid.

Chapter 7

1. Juliet Clutton-Brock, *Horse Power: A History of the Horse and the Donkey* (Cambridge, MA: Harvard University Press, 1992).
2. Arlene Weintraub, "Crossing the Gene Barrier," *Business Week,* January 16, 2006, pp. 72–80.
3. Jason Robert and Francoise Baylis, "Crossing Species Boundaries," *American Journal of Bioethics* 3, no. 3, (2003): 1–13; and Mark Greene, et al., "Moral Issues of Human-Non-Human Primate Neural Grafting," *Science,* 309, no. 5733 (2005): 385–386.
4. Jamie Shreeve, "The Other Stem-Cell Debate," *New York Times,* April 10, 2005.
5. *Homer's Iliad,* translation by Denison Bingham Hull, (Scottsdale, AZ: D. B. Hull, 1982).
6. Morgan Thomas, "Monsters of Our Minds," *Courier Mail,* July 9, 2003, 22.
7. *Pittsburgh Post-Gazette,* "Monsters on the Loose; Why Chimeras are So Very Much in the News These Days," May 9, 2005, B-7.
8. Rick Weiss, "Cloning Yields Human-Rabbit Hybrid Embryo," *Washington Post,* August 14, 2003, A4.
9. Laura Tangley, "A Therapy for Cat Allergies, Thanks to Mice," *New York Times,* April 5, 2005, 6.
10. Carolyn Johnson, "From Myth to Reality: Scientists Can Create Animals with the Cells of Other Species, but are these Chimeras Medical Marvels or High-Tech Monsters," *Boston Globe,* April 19, 2005, D1.
11. Rick Weiss, "Human Brain Cells Aare Grown in Mice," *Washington Post,* December 13, 2005, A3.
12. *Pittsburgh Post-Gazette,* "Monsters on the Loose."
13. Maryann Mott, "Animal-Human Hybrids Spark Controversy," *National Geographic News,* January 25, 2005.
14. Ibid.
15. Arlene Weintraub, "What's Ethical and What Isn't?" *Business Week,* January 16, 2006, 76.
16. Ibid.
17. Johnson, "From Myth to Reality."
18. *Jim Lehrer News Hour Transcript,* "Extended Interview: Hank Greely," July 2005.

19. National Academies of Science, *Guidelines for Human Embryonic Stem Cell Research* (Washington, D.C.: The National Academies Press, 2005).
20. Ibid.
21. *Bulletin's Frontrunner,* "Brownback Vows to Block Patents on Cloned Humans," May 21, 2002, 1.
22. *Jim Lehrer News Hour Transcript,* "Extended Interview: Irving Weissman," July 2005.
23. Mott, "Animal-Human Hybrids Spark Controversy."
24. Roger Highfield, "Stem Cell Researchers Plan to Create Rabbit-Human Embryos," *London Daily Telegraph,* January 13, 2006, 10.
25. Ibid.
26. Ian Johnston, "Embryos Key to Beating Genetic Killers," *The Scotsman,* January 13, 2006, 2.
27. Ibid.
28. Ibid.
29. Marie Woolf, "Record Number of GM Animals Being Bred in UK Laboratories," *London Independent,* December 4, 2005, 14.
30. Mark Henderson and Adam Sage, "Scientists Create Chickens with Teeth," *Ottawa Citizen,* June 4, 2003, A11.
31. Francis Fukayama, "The Clone Traders: Advances in Human Biotechnology Have Highlighted the Need for International Regulation," *Financial Times,* May 18, 2002 1.
32. *Pharma Marketletter,* "Moratorium in Xenotransplantation Urged," January 28, 1998.
33. Weintraub, "Crossing the Gene Barrier," *New Zealand Press Association,* "USA Company Engineering NZ Goats to Produce Blood Clotting Drug," January 13, 2006, and Stephen Heuser, "GTC Gets Surprise Boost from EU," *Boston Globe,* June 3, 2006, F7.
34. Emma Ross, "Fertility Society Denounces Creation of Human Chimera Embryo," *Associated Press Worldstream,* July 2, 2003.
35. Fukayama, "The Clone Traders."
36. Christopher Chapple, *Nonviolence to Animals, Earth, and Self in Asian Traditions* (Albany: State University of New York Press, 1993).
37. Weiss, "Cloning Yields Human-Rabbit Hybrid Embryo."
38. Dennis Normile and Charles Mann, "Asia Jockeys for Stem Cell Lead," *Science* 307, no. 5710 (2005): 660–664.
39. Weiss, "Cloning Yields Human-Rabbit Hybrid Embryo."
40. John von Radowitz, "Scientists 'Create Hybrid Human-Rabbit Embryos,'" *Press Association,* August 14, 2003.
41. Normile and Mann, "Asia Jockeys for Stem Cell Lead."
42. *The Times of India,* "Clinical Trials of Pig Organ Transplant in 2 Years," February 7, 2000.
43. *The Economist,* "Xeno's Paradox," December 21, 1996.
44. *Business Line,* "Human Kidneys Grown in Mice," December 25, 2002.

45. Dennis Normile, "Japan: Human Cloning Ban Allows Some Research," *Science* 290, no. 5498, (2000): 1872; and *Jiji Press Ticker Service,* "Japan to Approve Creation of Animal-Human Embryo," November 6, 2001, 1.
46. Michael Janofsky, "11 Indicted in Cases of Environmental Sabotage," *New York Times,* January 21, 2006, A1.
47. *Jim Lehrer News Hour Transcript,* "Extended Interview: Eugene Redmond," July 2005.
48. Shreeve, "The Other Stem-Cell Debate."
49. Ibid.
50. *Lehrer News Hour Transcript,* "Interview: Hank Greely."
51. Ibid.

Chapter 8

1. Nicole Gray, "Sixth Annual Report of the World's Top 50 Pharma Companies: Untying the Gordian Knot," *Pharm Exec,* May 2005.
2. N. Suresh, "Global Biotech Crosses $60-Billion Mark," *BioSpectrum,* May 12, 2006.
3. Justin Gillis, "Biotech's Gains Again Outstrip Drug Giants," *Washington Post,* April 12, 2006, D1.
4. Andrew Pollack, "Signs That Biotech Has a Healthier Future," *New York Times,* April 4, 2006, C4.
5. Andrew Pollack, "U.S. Finance Pulls Biotech Across Seas," *New York Times,* July 12, 2006, C1.
6. Ray Moynihan and Alan Cassels, *Selling Sickness: How the World's Biggest Pharmaceutical Companies are Turning Us All into Patients* (New York: Nation Books, 2005).
7. "Biotechnology Industry Organization Milestones 2004," undated report online at www.bio.org/speeches/pubs/milestone04/business.asp.
8. Peter Gwynne and Guy Page, "Biotechnology: Personal Portraits of an Evolving Industry," *Science* 287, no. 5461 (2000).
9. *Wall Street Reporter,* "Successful New Models for Biotech Big Pharma Partnership," June 21, 2006.
10. Jane Friedman, "New Report Shows Biotech Companies Bring More New Drugs to Market than Big Pharmaceuticals," *DigitalJournal.Com,* April 21, 2006.
11. Gwynne and Page, "Biotechnology: Personal Portraits."
12. Daniel Levine, "Big Pharmas Agreeing to Big Deals with Biotech Firms," *San Jose Business Journal,* June 20, 2003.
13. *DrugResearcher.com,* "Falling Innovation Levels Fuel Pharma/Biotech Collaborations," May 10, 2006.
14. Nicholas Wade, "Scientists Draft Rules on Ethics For Stem Cells," *New York Times,* April 27, 2005, 1.

15. Lee Romney, "Out-of-State Facility Demands Part of Stem Cell Research Royalties," *Los Angeles Times*, March 30, 2006, B3.
16. Ibid.
17. Ed Vulliamy, "Medicine and Ethics: The Cell Struggle," *Guardian Newspaper Observer*, July 8, 2001, 18.
18. Marcia Angell, *The Truth About the Drug Companies* (New York: Random House, 2004); Jerome Kassirer, *On the Take* (New York: Oxford University Press, 2005); Moynihan and Cassels, *Selling Sickness;* and Jerry Avorn, *Powerful Medicines* (New York: Alfred A. Knopf, 2004).
19. Angell, *The Truth About the Drug Companies.*
20. Ibid.
21. Wanda Hamilton, "Big Pharmaceuticals, Big Money," www.forces.org, August 24, 2001.
22. Shankar Vedantam and Marc Kaufman, "Doctors Influenced By Mention of Drug Ads," *Washington Post*, April 27, 2005, A1.
23. Arthur Caplan, "Indicting Big Pharma," *American Scientist*, January–February 2005.
24. Jim Drinkard, "Drugmakers Go Furthest to Sway Congress," *USA Today*, April 26, 2005, 1B.
25. Sonia Shah, *The Body Hunters: How the Drug Industry Tests its Products on the World's Poorest Patients* (New York: New Press, 2006).
26. Kelly Hearn, "Big Pharma's Deadly Experiments," *AlterNet*, June 9, 2006 online at www.alternet.org.
27. Jeanne DiGrazio, "Patenting Human Genes," *Science and Technology Newsletter*, Bryn Mawr College, April 2002.
28. Jonathan Eaton, Samuel Kortum, and Josh Lerner, "International Patenting and the European Patent Office: A Quantitative Assessment," *Patents, Innovation and Economic Performance, OECD Proceedings, 2004*, (OECD, 2004), 27–52, http://www.oecd.org/dataoecd/12/4/33793381.pdf, unpublished paper, August 2003.
29. Per Botolf Maurseth and Bart Verspagen, "Knowledge Spillovers in Europe: A Patent Citations Analysis," *Scandinavian Journal of Economics* 104, no. 4 (2002): 531–545.
30. Zhiqi Chen and Alison McDermott, "International Comparisons of Biotechnology Policies," *Journal of Consumer Policy* 21 (1998): 527–550.
31. Beppi Crosariol, "Why Laws Really Love Biotech," *Toronto Globe and Mail*, June 7, 2006.
32. Peter Drahos, "Developing Countries and International Intellectual Property Standard-Setting," *Journal of World Intellectual Property* 5, no. 5 (2002): 765–789.
33. Jose Cortina, "International Patent Laws Affect RTP Biotech Industry," Daniels, Daniels, and Verdonik, undated, http://www.d2vlaw.com/articles.htm.
34. Paroma Basu, "International Patent Law—Boon or Bane of Biotech?" *Nature Biotechnology* 23 (2005): 13–15.
35. Ibid.

36. Mariko Sakakibara and Lee Branstetter, "Do Stronger Patents Induce More Innovation? Evidence from the 1988 Japanese Patent Law Reforms" (NBER Working Paper Series, April 1999, unpublished paper).
37. Edwin Lai, "International Intellectual Property Rights Protection and the Rate of Product Innovation," *Journal of Development Economics* 55 (1998): 133–153.
38. Catherine Arnst, "Biotech, Finally," *Business Week,* June 13, 2005.
39. Ibid.
40. Ibid.
41. Jean-Francois Tremblay, "India Emerges in Drug Services," *Chemical and Engineering News* 83, no. 5 (2005): 15–17.
42. Calum MacLeod, "China Makes Ultimate Punishment Mobile," *USA Today,* June 15, 2006, 8A.
43. Ibid.
44. January Payne, "Still, A Cruel Diagnosis: In King's Death, a Lesson in Ovarian Cancer's Deadliness," *Washington Post,* February 14, 2006, F1.

Chapter 9

1. Ambuj Sagar, Arthur Daemmrich, and Mona Ashiva, "The Tragedy of the Commoners: Biotechnology and its Publics," *Nature* 18 (January 2000).
2. A. Bonnicksen, *Crafting a Cloning Policy: From Dolly to Stem Cells* (Washington, D.C.: Georgetown University Press, 2002).
3. T. Caulfield, L. Knowles, and E. Meslin, "Law and Policy in the Era of Reproductive Genetics," *Journal of Medical Ethics* 30 (2004): 414–417.
4. Michele Curtis, "Cloning and Stem Cells: Processes, Politics, and Policy," *Current Women's Health Reports* 3 (2003): 492–500.
5. Caulfield, Knowles, and Meslin, "Law and Policy."
6. Paul Quirk, *Industry Influence in Federal Regulatory Agencies* (Princeton, NJ: Princeton University Press, 1981); and Martha Derthick and Paul Quirk, *The Politics of Deregulation* (Washington, D.C.: Brookings Institution Press, 1985).
7. Derek Morgan, "Ethics, Economics and the Exotic: The Early Career of the HFEA," *Health Care Analysis* 12, no. 1 (2004): 7–26.
8. Sagar, Daemmrich, and Ashiva, "The Tragedy of the Commoners."
9. Helen Lawton Smith, "Regulating Science and Technology: The Case of the UK Biotechnology Industry," *Law and Policy* 27, no. 1 (2005): 189–212.
10. Ibid.
11. Brian Salter and Mavis Jones, "Regulating Human Genetics: The Changing Politics of Biotechnology Governance in the European Union," *Health, Risk & Society* 4, no. 3 (2002): 325–340.
12. Caulfield, Knowles, and Meslin, "Law and Policy."
13. Matthew Nisbet, "Public Opinion about Stem Cell Research and Human Cloning," *Public Opinion Quarterly* 68, no. 1 (2004): 131–154.
14. National Science Foundation, "Survey of Public Attitudes Toward and Understanding of Science and Technology," unpublished, 2001 report.

15. European Commission Special EuroBarometer 224, "Europeans, Science & Technology," unpublished, June 2005 report.
16. Ibid.
17. National Science Foundation, "Survey of Public Attitudes."
18. Japan National Institute of Science and Technology Policy, "The 2001 Survey of Public Attitudes Toward and Understanding of Science & Technology in Japan," unpublished, January 2002 report.
19. Nisbet, "Public Opinion."
20. Martin Bauer, John Durant, and Geoffrey Evans, "European Public Perceptions of Science," *International Journal of Public Opinion* 6, no. 2 (1994): 163–186.
21. National Science Board, "Information Sources, Interest, and Perceived Knowledge," unpublished, 2004 report.
22. *Science,* "Europeans Apathetic Toward Science," December 10, 2001.

Bibliography

Chapter 1 The Globalization of Innovation

Andreas, Peter, and Thomas Biersteker, eds. *The Rebordering of North America.* New York: Routledge, 2003.

Blumenstyk, Goldie. "Colleges Cash In on Commercial Activity." *Chronicle of Higher Education,* December 2, 2005, A25–26.

———. "Turning Research—Slowly—Into Riches: Technology Transfer Gains a Foothold at European Universities." *Chronicle of Higher Education,* October 7, 2005, A44–45.

Bok, Derek. *Universities in the Marketplace: The Commercialization of Higher Education.* Princeton, NJ: Princeton University Press, 2003.

Cerny, Philip. "Globalization and the Erosion of Democracy." *European Journal of Political Research* 36, no. 1 (1999): 1–26.

Doner, Richard. "Limits of State Strength: Toward an Institutionalist View of Economic Development." *World Politics* 44, no. 3 (1992): 398–431.

Drezner, Daniel. "The Global Governance of the Internet: Bringing the State Back In." *Political Science Quarterly* 119 (Fall 2004): 477–98.

Ellstrand, Norman. *Dangerous Liaisons? When Cultivated Plants Mate with Their Wild Relatives.* Baltimore: Johns Hopkins University Press, 2003.

Evans, Peter. *Embedded Autonomy: States and Industrial Transformation.* Princeton, NJ: Princeton University Press, 1995.

Fagerberg, Jan. "Technology and International Differences in Growth Rates." *Journal of Economic Literature* 32, no. 3 (1994): 1147–75.

Friedman, Thomas. *The Lexus and the Olive Tree.* New and expanded edition, New York: Anchor Books, 2000.

Haas, Peter. "Introduction: Epistemic Communities and International Policy Coordination." *International Organization* 46, (Spring 1992): 1–35.

Hauffler, Virginia. *A Public Role for the Private Sector.* Washington, D.C.: Carnegie Endowment for International Peace, 2001.

Huntington, Samuel. *The Clash of Civilizations and the Remaking of World Order.* New York: Simon & Schuster, 1996.

Jasanoff, Sheila. *Designs on Nature: Science and Democracy in Europe and the United States.* Princeton, NJ: Princeton University Press, 2005.

Kimball, Paul. "Globalization, the U.S. Economy & The Imperative of Innovation." Presentation at Brown University, March 16, 2005.

Kohli, Atul. *Democracy and Discontent: India's Growing Crisis of Governability.* New York: Cambridge University Press, 1990.

Krimsky, Sheldon. *Science in the Private Interest.* Lanham, MD: Rowman & Littlefield, 2003.

Luke, Timothy, and Gearoid Tuanthail. "The Fraying Modern Map: Failed States and Contraband Capitalism." unpublished paper, undated.

National Science Board. "Science and Engineering Indicators 2004." Washington, D.C.: National Science Foundation, 2004.

Organisation for Economic Co-Operation and Development. *Compendium of Patent Statistics, 2004.* Paris: OECD, 2004.

———. *Science and Technology Statistical Compendium, 2004.* Paris: OECD, 2004.

Petersen, Melody. "A Conversation with Sheldon Krimsky: Uncoupling Campus and Company." *New York Times,* September 23, 2003, F2.

Pevehouse, Jon. "Democratization, Credible Commitments, and Joining International Organizations." In *Locating the Proper Authorities: The Interaction of Domestic and International Institutions,* edited by Daniel Drezner, 25–48. Ann Arbor: University of Michigan Press, 2003.

Rai, Arti, and Rebecca Eisenberg. "Bayh-Dole Reform and the Progress of Biomedicine." *American Scientist.* 91 (2003): 53.

Rodrik, Dani. *Has Globalization Gone Too Far?* Washington, D.C.: Institute for International Economics, 1997.

Rosecrance, Richard. *The Rise of the Virtual State.* New York: Basic Books, 1999.

Saxenian, AnnaLee. *Regional Advantage: Culture and Competition in Silicon Valley and Route 128.* Cambridge, MA: Harvard University Press, 1994.

Scholte, Jan Aart. *Globalization: A Critical Introduction.* New York: St. Martins Press, 2000.

Segal, Adam. *Digital Dragons.* Ithaca, NY: Cornell University Press, 2003.

Slotten, Hugh. "Satellite Communications, Globalization, and the Cold War." *Technology and Culture.* 43 (April 2002): 315–50.

Stiglitz, Joseph. *Globalization and Its Discontents.* New York: W. W. Norton, 2003.

Strange, Susan. *The Retreat of the State: The Diffusion of Power in the World Economy.* New York: Cambridge University Press, 1996.

Tonelson, Alan. *The Race to the Bottom,* Boulder, CO: Westview Press, 2000.

Wade, Robert. *Governing the Market: Economic Theory and the Role of Government in East Asian Industrialization.* Princeton, NJ: Princeton University Press, 1990.

Wapner, Paul. "Politics Beyond the State: Environmental Activism and World Civic Politics." *World Politics.* 47 (April 1995): 311–40.

Washburn, Jennifer. *University, Inc.: The Corporate Corruption of Higher Education,* New York: Basic Books, 2005.

Wilmut, Ian, Keith Campbell, and Colin Tudge. *The Second Creation: The Age of Biological Control by the Scientists that Cloned Dolly.* London: Headline, 2000.

Yergin, Daniel, and Joseph Stanislaw. *The Commanding Heights: The Battle Between Government and the Marketplace That is Remaking the Modern World.* New York: Simon & Schuster, 1998.

Zysman, John. *Governments, Markets, and Growth.* Ithaca, NY: Cornell University Press, 1983.

Chapter 2 Science-Industry Collaboration

Alistair, Brett, David Gibson, and Raymond Smilor, eds. *University Spin-off Companies: Economic Development, Faculty Entrepreneurs, and Technology Transfer.* Savage, MD: Rowman & Littlefield, 1991. xxii.

Anselin, Luc, Attila Varga, and Zoltan Acs. "Local Geographic Spillovers Between University Research and High Technology Innovations." *Journal of Urban Economics.* 42, no. 3 (1997): 422–48.

Association of University Technology Managers. *AUTM Licensing Survey: FY 2002.* Northbrook, IL: AUTM, 2003.

Barboza, David. "China's Problem with 'Anti-Pest' Rice." *New York Times,* April 16, 2005, B1.

Butler, Declan. "French Scientists Offered Time to Set up Companies." *Nature* 397, no. 6716 (1999): 187.

Derthick, Martha, and Paul Quirk. *The Politics of Deregulation.* Washington, D.C.: Brookings Institution, 1985.

Gibson, David, and Raymond Smilor. "The Role of the Research University in Creating and Sustaining the U.S. Technopolis." In *University Spin-off Companies: Economic Development, Faculty Entrepreneurs, and Technology Transfer* edited by Alistair Brett, David Gibson, and Raymond Smilor, 31–70. Savage, MD: Rowman & Littlefield, 1991.

Harris, Kenneth. *Thatcher.* Boston, MA: Little Brown, 1988.

Jasanoff, Sheila. *Designs on Nature: Science and Democracy in Europe and the United States.* Princeton, NJ: Princeton University Press, 2005.

Jasanoff, Sheila. *The Fifth Branch: Science Advisers as Policymakers.* Cambridge, MA: Harvard University Press, 1990.

Johnson, Haynes. *Sleepwalking Through History: America in the Reagan Years.* New York: Norton, 2003.

Krimsky, Sheldon, L. Rothenberg, P. Stott, and G. Kyle. "Financial Interests of Authors in Scientific Journals." *Science and Engineering Ethics* 2, no. 4 (1996): 395–410.

Krimsky, Sheldon. *Science in the Private Interest.* Oxford: Rowman & Littlefield, 2003.

Laville, Sandra, and Duncan Campbell. "Animal Rights Extremists in Arson Spree." *Guardian,* July 1, 2005, 10.

Link, Albert. *A Generosity of Spirit: The Early History of the Research Triangle Park.* Research Triangle Park, NC: Research Triangle Foundation of North Carolina, 1995.

Luger, Michael, and Harvey Goldstein. *Technology in the Garden: Research Parks and Regional Economic Development.* Chapel Hill, NC: University of North Carolina Press, 1991.

McLaren, Anne. "The Decade of the Sheep." *Nature* 403, no. 6769 (2000): 479.

Nadkarni, Namrata. "US Exports of GM Modified Corn Hit a New Setback: Japan May Ship Back or Destroy Unauthorised Cargo." *Lloyd's List,* June 28, 2005, 4.

National Science Board. *Science and Engineering Indicators, 2004.* Arlington, Virginia: National Science Foundation, 2004.

Normile, Dennis, and Charles Mann. "Asia Jockeys for Stem Cell Lead." *Science* 307, no. 5710 (2005): 662.

Normile, Dennis. "Can Money Turn Singapore into a Biotech Juggernaut?" *Science* 297, no. 5586 (2002): 488.

Organisation for Economic Co-Operation and Development. *Main Science and Technology Indicators, 2002.* Paris: OECD, 2002.

Overland, Martha. "A Tale of 2 Countries: Singapore's Regeneration." *Chronicle of Higher Education,* November 11, 2005, A42–46.

Peterson, Melody. "A Conversation with Sheldon Krimsky: Uncoupling Campus and Company." *New York Times,* September 23, 2003, F2.

Riddell, Peter. *The Thatcher Era and Its Legacy.* Oxford: Blackwell, 1991.

Rothenberg, Lawrence. *Regulation, Organizations, and Politics.* Ann Arbor: University of Michigan Press, 1994.

Savas, Emanuel. *Privatization and Public-Private Partnerships.* New York: Chatham House, 2000.

Savoie, Donald. *Thatcher, Reagan, Mulroney: In Search of a New Bureaucracy.* Pittsburgh, PA: University of Pittsburgh Press, 1994.

Science. "Biotechnology Start-ups in Singapore: Inspiring Future Entrepreneurs." March 22, 2002.

Shreeve, Jamie. "The Other Stem-Cell Debate" *New York Times Magazine.* April 10, 2005, 44.

Stockman, David. *The Triumph of Politics: How the Reagan Revolution Failed.* New York: Harper & Row, 1986.

Washburn, Jennifer. *University, Inc.: The Corporate Corruption of Higher Education.* New York: Basic Books, 2005.

West, Darrell M. *Congress and Economic Policymaking.* Pittsburgh, PA: University of Pittsburgh Press, 1987.

Yonhap News Agency. "Korean Air Gives Cloning Expert 10-Year Free Flights," June 3, 2005, 1.

Chapter 3 In-Vitro Fertilization

Abbott, Alison. "Germany's Past Still Casts a Long Shadow." *Nature.* 389, no. 6652 (1997): 660.

Allen, Mike, and Rick Weiss. "Bush Rejects Stem Cell Compromise." *Washington Post,* May 26, 2005, A2.

Andersen, A., L. Nyboe, Gianaroli, R. Felberbaum, J. de Mouzon, and K. G. Nygren. "Assisted Reproductive Technology in Europe, 2001." *Human Reproduction* 20, no. 5 (2005): 1158–76.

Arie, Sophie. "In Europe, Italy Now a Guardian of Embryo Rights." *Christian Science Monitor,* June 14, 2005, 1.

Associated Press. "Swiss Voters Reject New Initiatives," December 3, 2000.

Bahnsen, Ulrich. "Swiss to Vote on Ban on In Vitro Fertilization." *Nature* 396, no. 6707 (1998): 105.

Barnhart, Michael. "Nature, Nurture, and No-Self: Bioengineering and Buddhist Values." *Journal of Buddhist Ethics* 7 (2000): 126–144.

Bharadwaj, Aditya. "How Some Indian Baby Makers are Made: Media Narratives and Assisted Conception in India." *Anthropology and Medicine* 7, no. 1 (2000): 63–78.

Bilefsky, Dan. "Court Rules Couple Must Agree on Use of Embryos." *International Herald-Tribune,* March 8, 2006, 1.

Bleiklie, Ivar. "Governing Assisted Reproductive Technology: The Case of Norway." Paper prepared for delivery at the annual meeting of the American Political Science Association, San Francisco, August 30–September 2, 2001.

Boggio, Andrea. "Italy Enacts New Law on Medically Assisted Reproduction." *Human Reproduction* 20, no. 5 (2005): 1153–57.

Boseley, Sarah. "Fertility Debate: Public Asked to Help Rewrite IVF Law." *The Guardian,* August 17, 2005, 4.

Centers for Disease Control and Prevention. "Assisted Reproductive Technology." www.cdc.gov/ART/index.htm.

Cohen, Cynthia. "Unmanaged Care: The Need to Regulate New Reproductive Technologies in the United States." *Bioethics* 11, no. 3 & 4 (1997): 348–65.

Congregation for the Doctrine of the Faith. "Instruction on Respect for Human Life in its Origin and on the Dignity of Procreation Replies to Certain Questions of the Day," online at www.vatican.va, February 22, 1987.

Donnellan, Eithne. "Couples Going Abroad for Tests on Embryos." *Irish Times,* September 16, 2005, 3.

Doring, Ole. "China's Struggle for Practical Regulations in Medical Ethics." *Nature* 4 (March 2003), 233–39.

Dunson, David, Donna Baird, and Bernardo Colombo. "Increased Infertility With Age in Men and Women." *Obstetrics & Gynecology* 103 (2004): 51–56.

Heitman, Elizabeth. "Social and Ethical Aspects of In Vitro Fertilization." *International Journal of Technology Assessment in Health Care* 15, no. 1 (1999): 22–35.

InterNational Council on Infertility Information Dissemination. "A History of IVF Statistics." www.inciid.org.

Irish Times. "Public Divided on Infertility Treatment Ethics." May 13, 2005, 9.

Jasanoff, Sheila. *The Fifth Branch: Science Advisers as Policymakers.* Cambridge, MA: Harvard University Press, 1990.

Johnson, Martin. "Should the Use of Assisted Reproduction Techniques Be Deregulated?" *Human Reproduction* 13, no. 7 (1998): 1769–76.

Karande, Vishvanath, Alan Korn, Randy Morris, Ramaa Rao, Martin Balin, John Rinehart, Karen Dohn, and Norbert Gleicher. "Prospective Randomized Trail Comparing the Outcome and Cost of In Vitro Fertilization with that of a Traditional Treatment Algorithm as First-Line Therapy for Couples with Infertility." *Fertility and Sterility* 71, no. 3 (1999): 468–75.

Kiefer, Heather Mason. "The Birth of In Vitro Fertilization." *The Gallup Poll,* August 5, 2003.

Kovacs, Gabor, Gary Morgan, Carl Wood, Catherine Forbes, and Donna Howlett. "Community Attitudes to Assisted Reproductive Technology: a 20-Year Trend." *Medical Journal of Australia* 179, (November 17, 2003): 536–38.

Langdridge, Darren, and Eric Blyth. "Regulation of Assisted Conception Services in Europe." *Journal of Social Welfare and Family Law* 23, no. 1 (2001): 45–64.

Le Moyne College/Zogby International. "Contemporary Catholic Trends." November 16, 2001. www.Zogby.com.

Macer, Darryl, J. Azariah, and P. Srinives. "Attitudes to Biotechnology in Asia." *International Journal of Biotechnology* 2 (2000): 313–32.

Macer, Darryl. "Perception of Risks and Benefits of In Vitro Fertilization, Genetic Engineering and Biotechnology." *Social Science and Medicine* 38 (1994): 23–33.

Merriam-Webster's Collegiate Dictionary. http://search.eb.com/dictionary.

Mishra, Pankaj. "How India Reconciles Hindu Values and Biotech." *New York Times,* August 21, 2005, 4.

Mitchell, Susan. "We're having an IVF Baby." *Financial Times,* September 19, 2004.

Moses, Lyria Bennett. "Understanding Legal Responses to Technological Change: The Example of In Vitro Fertilization." *Minnesota Intellectual Property Review,* (May 2005).

Mudur, Ganapati. "India Considers Government Agency to License Infertility Clinics." *British Medical Journal* 325 (September 14, 2002): 564.

Qiu, Ren-Zong. "Sociocultural Dimensions of Infertility and Assisted Reproduction in the Far East." Online at www.who.int/reproductive-health/infertility/12.pdf, undated.

Ramjoue, Celina. "Assisted Reproductive Technology Policy in Italy." Paper presented at the ECPR Joint Sessions Workshop, March 22–27, 2002, Turin, Italy.

Reid, Liam, and Carol Coulter. "Regulatory Body for Fertility Treatment Urged." *Irish Times,* March 13, 2006, 8.

Reynolds, Meredith, and Laura Schieve. "Insurance Coverage and Outcomes of in Vitro Fertilization." *New England Journal of Medicine* 348, no. 10 (2003): 958–59.

Robertson, John. "Assisted Reproduction in Germany and the United States." *Berkeley Electronic Press,* Paper 226 (April 1, 2004): 1–46.

Robison, Jennifer. "Infertile Women Quest for Family." *The Gallup Poll,* May 21, 2002.

Rosenthal, Elisabeth, and Elisabetta Povoledo. "Vote on Fertility Law Fires Passions in Italy." *New York Times,* June 11, 2005, 7.

Schenker, Joseph. "Assisted Reproduction Practice in Europe: Legal and Ethical Aspects." *Human Reproduction Update* 3, no. 2 (1997): 173–84.

Senat, Marie-Victoire, Pieerre-Yves Ancel, Marie-Helene Bouvier-Colle, and Gerard Breart. "How Does Multiple Pregnancy Affect Maternal Mortality and Morbidity?" *Clinical Obstetrics and Gynecology* 41, no. 1 (1998): 79–83.

Spar, Debora. *The Baby Business: How Money, Science, and Politics Drive the Commerce of Conception.* Cambridge, MA: Harvard Business School Press, 2006.

Strickler, Jennifer. "The New Reproductive Technology." *Sociology of Health & Illness* 14, no. 1 (1992): 111–32.

20th Century History. "First Test-Tube Baby—Louise Brown." www.history1900s.about.com.

Warner, Carol. "Research Description." www.biology.neu.edu/faculty03/warner03.html.

Weiss, Rick. "Babies in Limbo: Laws Outpaced by Fertility Advances." *Washington Post,* February 8, 1998, A1.

Widge, Anjali. "Sociocultural Attitudes Towards Infertility and Assisted Reproduction in India." Online at www.who.int/reeproductive-health/infertility/11.pdf, undated.

Yu Ng, Ernest Hung, Athena Liu, Celia Chan, Cecilia Lai Wan Chan, William Shu Biu Yeung, and Pak Chung Ho. "Regulating Reproductive Technology in Hong Kong." *Journal of Assisted Reproduction and Genetics* 20, no. 7 (2003): 281–86.

Zhang, Ya. "The Mutual Interaction Between Genetic Information Technology and Societal Factors in China." Unpublished paper, School of IST, Penn State University, 2003.

Chapter 4 Genetically Modified Food

Abbott, Allison, and Burkhardt Roeper. "Germany Seeks 'Non-Modified' Food Label." *Nature* 391, no. 6670 (1998):828.

Agence France Presse. "China Planning Large-Scale Introduction of Genetically Engineered Rice." November 30, 2004.

Alden, Edward, and Jeremy Grant. "WTO Rules Against Europe in GM Food Case." *Financial Times,* February 8, 2006, 6.

Barboza, David. "China's Problem With 'Anti-Pest' Rice." *New York Times,* April 16, 2005, B1.

———. "Modified Rice May Benefit China Farms, Study Shows." *New York Times,* May 3, 2005, C8.

Bernauer, Thomas, and Erika Meins. "Scientific Revolution Meets Policy and the Market: Explaining Cross-National Differences in Agricultural Biotechnology Regulation." CIES Discussion Paper No. 144, November 2001.

Burton, Michael, Dan Rigby, Trevor Young, and Sallie James. "Consumer Attitudes to Genetically Modified Organisms in Food in the UK." *European Review of Agricultural Economics* 28, no. 4 (2001): 479–98.

Business Daily Update. "GM Rice May Soon Be Commercialized." January 28, 2005, section 28.

BBC Monitoring International Reports. "China Ratifies GMO Trade Protocol." May 19, 2005.

Carter, Colin, and Guillaume Gruere. "International Approaches to the Labeling of Genetically Modified Foods." *Choices* (Second Quarter 2003): 1–4.

Dela Cruz, Roderick. "UNCTAD Calls on RP, Other Countries to Balance Impact of GMO." *Manila Standard,* May 17, 2005.

Dyer, Geoff. "Controversy Grows Over China's Biotech Crops." *Financial Times,* June 24, 2005, 14.

Europe Agri. "Genetic Engineering: Madrid Plans Safety Distances Between Traditional and GM Crops." December 21, 2004, 1.

Fedoroff, Nina, and Nancy Brown. *Mendel in the Kitchen: A Scientist's View of Genetically Modified Foods.* Washington, D.C.: Joseph Henry Press, 2004.

Feffer, John. "Asia Holds Key to Future of GM Food." *Korea Herald,* December 6, 2004.

Fulton, Murray, and Konstantinos Giannakas. "Agricultural Biotechnology and Industry Structure." *AgBioForum* 4, no. 2 (2001): 137–51.

Gaskell, George, Martin Bauer, John Durant, and Nicholas Allum. "Worlds Apart? The Reception of Genetically Modified Foods in Europe and the U.S." *Science* 285 no. 5426 (1999): 384–87.

Genetically Engineered Organisms Public Issues Education Project. "GE Foods in the Market." Ithaca, NY: Cornell Cooperative Extension, 2003.

Hallman, William, Carl Hebden, Helen Aquino, Cara Cuite, and John Lang. "Public Perceptions of Genetically Modified Foods." New Brunswick, NJ: Food Policy Institute, October 2003.

Hautea, Randy, and Margarita Escaler. "Plant Biotechnology in Asia." *AgBioForum* 7, no. 1 & 2, 2004, Article 1.

Hickman, Karma. "Biotech Industry Calls for GMO Education." *ANSA English Media Service,* March 14, 2005.

Hino, Akihiro. "Safety Assessment and Public Concerns for Genetically Modified Food Products: The Japanese Experience." *Toxicologic Pathology* 30, no. 1 (2002): 126–28.

Horsch, Rob, and Jill Montgomery. "Why We Partner: Collaborations Between the Private and Public Sectors for Food Security and Poverty Alleviation through Agricultural Biotechnology." *AgBioForum* 7, no. 1 & 2 (2004): 80–83.

Huang, Jikun, Qinfang Wang, and James Keeley. "Agricultural Biotechnology Policy Processes in China." Unpublished paper, Institute for Development Studies, University of Sussex, August 2001, http://www.ids.ac.uk/idS/KNOTS/Projects/biotech/pubsPolproc.html.

Inaba, Masakazu, and Darryl Macer. "Policy, Regulation and Attitudes towards Agricultural Biotechnology in Japan." *Journal of International Biotechnology Law* 1 (2004).

Jayaraman, K. S. "India Approves Use of Genetically Modified Crops, Despite Critics." *Nature* 397 no. 6716 (1999): 188.

Kolata, Gina. "Pork That's Good for the Heart May Be Possible with Cloning." *New York Times,* March 27, 2006, A1.

Lambrecht, Bill, and Deirdre Shesgreen. "Monsanto Lobbies to Keep the Status Quo for Gene-Altered Crops." *St. Louis Post-Dispatch,* September 11, 2005, A9.

Lei, Xiong. "China Could Be First Nation to Approve Sale of GM Rice." *Science* 306, no. 5701(2004): 1458–59.

Macer, Darryl, J. Azariah, and P. Srinives. "Attitudes to Biotechnology in Asia." *International Journal of Biotechnology* 2 (2000): 313–32.

Madkour, Magdy, Latha Nagarajan, and Clemen Gehlhar. "Private-Public Partnerships in Bringing BioTechnology from Laboratory to the Market Place: Comparison of Egypt and India." Paper presented at 6th International ICABR Conference, Ravello, Italy, July 11–14, 2002.

Marchant, Mary, Cheng Fang, and Baohui Song. "Issues on Adoption, Import Regulations, and Policies for Biotech Commodities in China with a Focus on Soybeans." *AgBioForum* 5, no. 4 (2002): 167–74.

Masood, Ehsan. "Royal Society Wants Genetics Watchdog." *Nature* 395, no. 6697 (1998): 5.

McDevitt, Thomas. *World Population Profile: 1998.* Washington, D.C.: U.S. Government Printing Office, 1999, A12.

Meller, Paul. "Europeans to Toughen Rules on Animal Feed From U.S." *New York Times,* April 13, 2005, C4.

———. "Europe Rejects Looser Labels for Genetically Altered Food." *New York Times,* September 9, 2004, 7.

Miller, Scott, and Joel Clark. "EU Orders Greece to Lift Gene-Altered Seed Ban." *Wall Street Journal,* January 11, 2006, A13.

Millstone, Erik, Eric Brunner, and Sue Mayer. "Beyond 'Substantial Equivalence.'" *Nature* 401, no. 6753 (1999): 525–6.

Nadkarni, Namrata. "US Exports of GM Modified Corn Hit a New Setback." *Lloyd's List,* June 28, 2005, 4.

New York Times. "Group Reports Increase in Biotech Harvest." January 13, 2005, C2.

Pew Initiative on Food and Biotechnology. "Recent Poll Findings." August 5–10, 2003 survey online at http://pewagbiotech.org.

Phillips, Peter, and Heather McNeill. "A Survey of National Labeling Policies for GM Foods." *AgBioForum* 3, no. 4 (2000): article 7.

Pollack, Andrew. "Lax Oversight Found in Tests of Gene-Altered Crops." *New York Times,* January 3, 2006, D2.

———. "Trade Ruling is Expected to Favor Biotech Food." *New York Times,* February 6, 2006, C6.

Randerson, James. "By the People, For the People: Genetically Modified Crops Don't Just Come from Big Corporations." *New Scientist* (February 19, 2005): 36.

Rosendal, Kristin. "Governing GMOs in the EU: A Deviant Case of Environmental Policy-Making?" *Global Environmental Politics* 5, no. 1 (2005): 82.

Safrin, Sabrina. "Treaties in Collision? The Biosafety Protocol and the World Trade Organization Agreements." *American Journal of International Law* 96 (2002): 606–28.

Scott, Alex. "Brussels Presses EU Members to Hasten GM Approvals." *Chemical Week,* March 30, 2005.

Sheldon, Ian. "Regulation of Biotechnology: Will We Ever 'Freely' Trade GMOs?" *European Review of Agricultural Economics* 29, no. 1 (2002): 155–76.

Taylor, Michael, and Jody Tick. "The StarLink Case: Issues for the Future." Unpublished paper prepared by Resources for the Future and the Pew Initiative on Food and Biotechnology, undated.

Turkish Daily News. "Turkey Working to Form Policy on Genetically Modified Organisms." May 25, 2005, 1.

U.S. Food and Drug Administration. "FDA's Policy for Foods Developed by Biotechnology." Unpublished 1995 report online at www.cfsan.fda.gov.

U.S. Food and Drug Administration. "Safety Assurance of Foods Derived by Modern Biotechnology in the United States." Unpublished July 1996 presentation online at www.cfsan.fda.gov.

Vogel, David, and Diahanna Lynch. "The Regulation of GMOs in Europe and the United States." Paper presented to the Council on Foreign Relations, New York, April 5, 2001.

Wall Street Journal. "Biotech Crops Grew at Slowest Pace Since 1996." January 12, 2006, A11.

Watanebe, Kazuo, Mohammad Taeb, and Haruko Okusu. "Japanese Controversies over Transgenic Crop Regulation," *Science* 305, no. 5690 (2004): 1572.

Williams, Nigel. "Agricultural Biotech Faces Backlash in Europe." *Science* 281, no. 5378(1998): 768–71.

Wright, Tom. "Swiss Voters Approve Ban on Genetically Altered Crops." *New York Times,* November 28, 2005, C2.

Chapter 5 Cloning

Becker, Gerhold. "Cloning Humans? The Chinese Debate and Why it Matters." *Eubios Journal of Asian and International Bioethics* 7 (1997): 175–78.

Bernier, Louise, and D. Gregoire. "Reproductive and Therapeutic Cloning, Germline Therapy, and Purchase of Gametes and Embryos: Comments on Canadian Legislation Governing Reproduction Technologies." *Journal of Medical Ethics* 30 (2004): 527–32.

Best, Steven, and Douglas Kellner. "Biotechnology, Ethics and the Politics of Cloning." *Democracy and Nature* 8, no. 3 (2002): 439–65.

Bruce, Donald. "Polly, Dolly, Megan, and Morag: A View from Edinburgh on Cloning and Genetic Engineering." *Philosophy and Technology* 3, no. 2 (1997): 37–52.

Dinc, Leyla. "Ethical Issues Regarding Human Cloning." *Nursing Ethics* 10, no. 3 (2003): 238–54.

Evans, John. "Cloning Adam's Rib: A Primer on Religious Responses to Cloning." Report prepared for the Pew Forum on Religion and Public Life, undated.

Fox, Cynthia. "Cloning Laws, Policies, and Attitudes Worldwide." *IEEE Engineering in Medicine and Biology* (March/April 2004): 55–61.

Frith, Michael. "Asian Nations Approach Cloning Consensus." *Nature Medicine* 9 (2003): 248.

Frommer, Frederic. "Dairy Industry, Consumers Skeptical of Cloned Cows." *Providence Journal,* July 13, 2005, E8.

Galli, Cesare, Roberto Duchi, Irina Lagutina, and Giovanna Lazzari. "A European Perspective on Animal Cloning and Government Regulation." *IEEE Engineering in Medicine and Biology* (March/April 2004): 52–54.

Highfield, Roger. "Embryo Cloning for Diabetes Research." *London Daily Telegraph,* April 20, 2005, 12.

Hopkins, Patrick. "Bad Copies: How Popular Media Represent Cloning as an Ethical Problem." *Hastings Center Report* 28, no: 2 (1998).

Indian Council of Medical Research. *Ethical Guidelines for Biomedical Research on Human Subjects.* New Delhi, India: ICMR, 2000.

Jaenisch, Rudolf, and Ian Wilmut. "Don't Clone Humans!" *Science* 291, no. 5513 (2001): 2552.

Kass, Leon. "The Wisdom of Repugnance: Why We Should Ban the Cloning of Humans." *New Republic* 216, no. 22 (1997): 17–23.

Klotzko, Arlene. *A Clone of Your Own? The Science and Ethics of Cloning* New York: Oxford University Press, 2004.

Lisker, Ruben. "Ethical and Legal Issues in Therapeutic Cloning and the Study of Stem Cells." *Archives of Medical Research* 34, no. 6 (2003): 607–11.

Lynch, Colum. "U.N. Backs Human Cloning Ban." *Washington Post,* March 9, 2005, A15.

Macer, Darryl. "Bioethics in Asia." in *Encyclopedia of the Human Genome.* London: Nature Macmillan, 2003.

Mahowald, Mary. "The President's Council on Bioethics, 2002–2004." *Perspectives on Biology and Medicine* 48 (2005): 159–71.

Maienschein, Jane. "What's in a Name: Embryos, Clones, and Stem Cells." *American Journal of Bioethics* 2 (2002): 12–19.

Messikomer, Carla, Renee Fox, and Judith Swazey. "The Presence and Influence of Religion in American Bioethics." *Perspectives in Biology and Medicine* 44, no. 4 (2001): 485–508.

Monastersky, Richard. "A Second Life for Cloning." *Chronicle of Higher Education,* February 3, 2006, A14–17.

Munro, Margaret. "S. Koreans Tailor Stem Cells to Fit Disease." *Montreal Gazette,* May 20, 2005, A25.

Nature. "China's Human-Cloning Policy Fudges Law on Cross-Species Fusions." 427, no. 6972 (2004): 278.

———. "UN Compromise Ends Human Cloning Debate with 'Non-Binding' Ban." 434, no. 7031 (2005): 264.

Nisbet, Matthew. "Public Opinion about Stem Cell Research and Human Cloning." *Public Opinion Quarterly* 68, no. 1 (2004): 131–54.

Parry, Sarah. "The Politics of Cloning: Mapping the Rhetorical Convergence of Embryos and Stem Cells in Parliamentary Debates." *New Genetics and Society* 22, no. 2 (2003): 145–68.

Pattinson, Shaun, and Timothy Caulfield. "Variations and Voids: The Regulation of Human Cloning Around the World." *BMC Medical Ethics* 5, no. 9 (2004): 10–18.

Pearson, Helen. "Cloning Success Marks Asian Nations as Scientific Tigers." *Nature* 427 no. 6976 (2004): 664.

President's Council on Bioethics. *Human Cloning and Human Dignity: An Ethical Inquiry.* Washington, D.C.: President's Council on Bioethics, 2002.

"*Public Opinion: Detailed Survey Results.*" January 2003 EOS Gallup Europe survey online at www.genetics-and-society.org/analysis/opinion/detailed.html.

Rosen, Elisabeth. "The Dolly-Dollar Dichotomy: Animal Cloning Restrictions and the Competitiveness of the European Biotech Industry." *Nordic Journal of International Law* 67 (1998): 423–30.

Saegusa, Asako. "Japan's Bioethics Debate Lags Behind Thinking in the West." *Nature,* Vol. 389, no. 6652 (1997): 661.

Salter, Brian, and Mavis Jones. "Regulating Human Genetics: The Changing Politics of Biotechnology Governance in the European Union." *Health, Risk & Society* 4, no. 3 (2002): 325–40.

Sanchez-Sweatman, L. R. "Reproductive Cloning and Human Health: An Ethical, International, and Nursing Perspective." *International Nursing Review* 47, no. 1 (2000): 28.

Soh-jung, Yoo. "Government to Establish Global Consortium for Cloning Research." *Korea Herald*, May 23, 2005.

Spurgeon, Brad. "France Bans Reproductive and Therapeutic Cloning." *British Medical Journal* 329, no. 7458 (2004): 130.

Stice, Steven. "Animal Cloning, Enhancements, and Genetic Selection." Paper presented at the 38th Annual Veterinary Conference, University of Georgia, April 6–8, 2001, http://www.georgiacenter.uga.edu/conferences/2001/Apr/06/vet_alumni.phtml.

Templeton, Sarah-Kate. "Dying Briton Pins Last Hope on Cloning." *London Sunday Times*, May 22, 2005, 12.

Webber, H. J. "New Horticultural and Agricultural Terms." *Science* 28 (1903): 501–3.

Wilmut, Ian, Keith Campbell, and Colin Tudge. *The Second Creation: Dolly and the Age of Biological Control.* New York: Farrar, Straus and Giroux, 2000.

Woloschak, Gayle. "Transplantation: Biomedical and Ethical Concerns Raised by the Cloning and Stem-Cell Debate." *Zygon* 38, no. 3 (2003): 699–704.

Chapter 6 Stem Cell Research

Akabayashi, Akira, and Brian Slingsby. "Biomedical Ethics in Japan." *Cambridge Quarterly of Healthcare Ethics* 12 (2003): 261–64.

Aldhous, Peter. "After the Gold Rush." *Nature* 434, no. 7034 (2005): 694–96.

Asia Africa Intelligence Wire. "Task Force to Seek Funds for Stem Cell Research." April 6, 2005.

Ayer, Allison. "Stem Cell Research: The Laws of Nations and a Proposal for International Guidelines." *Connecticut Journal of International Law* (Spring 2002): 35–49.

Beckmann, Jan. "On the German Debate on Human Embryonic Stem Cell Research." *Journal of Medicine and Philosophy* 29, no. 5 (2004): 603–21.

Benoit, Bertrand. "Schroeder in Call on Stem Cell Research." *Financial Times*, June 15, 2005, 9.

Bush, George. "I Have Given this Issue a Great Deal of Thought." *Science* 293, no. 5533 (2001): 1245.

Campbell, Angela. "Ethos and Economics: Examining the Rationale Underlying Stem Cell and Cloning Research Policies in the United States, Germany, and Japan." *American Journal of Law & Medicine* (2005): 77–86.

Capron, Alexander. "Stem Cells: Ethics, Law and Politics." *Biotechnology Law Report* 20, no. 5 (2001): 678–97.

Castellani, Federica. "Europe's Stem Cell Workers Pull Together." *Nature* 432, no. 7015 (2004): 260.

Doring, Ole. "Chinese Researchers Promote Biomedical Regulations." *Kennedy Institute of Ethics Journal* 14, no. 1 (2004): 39–46.

Dreifus, Claudia. "At Harvard's Stem Cell Center, the Barriers Run Deep and Wide." *New York Times,* January 24, 2006, D2.

Einhorn, Bruce. "A Cancer Treatment You Can't Get Here." *Business Week,* March 6, 2006, 88–90.

Harris, Sara. "Asian Pragmatism: Japan Has Set up a Legal Framework to Allow the Use and Generation of Human Embryonic Stem Cell Lines." *EMBO Reports* 3, no. 9 (2002): 816–17.

Hauskeller, Christine. "How Traditions of Ethical Reasoning and Institutional Processes Shape Stem Cell Research in Britain." *Journal of Medicine and Philosophy* 29, no. 5 (2004): 509–32.

Hindustan Times, "No Law, No Check on Stem Cell Clinics." March 18, 2005.

Human, Katy. "U.S. Researchers Recruited Labs in China, South Korea and England are Attracting Scientists Frustrated by the Lack of Stem Cell Funding." *Denver Post,* May 25, 2005, A21.

India Business Insight. "India Plans Fund for Stem Cell Research." April 7, 2005.

Jayaraman, K. S. "Indian Regulations Fail to Monitor Growing Stem Cell Use in Clinics." *Nature* 434, no. 7031 (2005): 259.

Juengst, Eric, and Michael Fossel. "The Ethics of Embryonic Stem Cells—Now and Forever, Cells Without End." *JAMA* 284, no. 24 (2000): 3180–84.

Kass, Leon. "The Wisdom of Repugnance." *New Republic* 216, no. 22 (1997): 17–23.

Kolata, Gina. "Koreans Report Ease in Cloning for Stem Cells." *New York Times,* May 20, 2005, A1.

Macer, Darryl. "Asian Approaches to Stem Cell Research and IP Protection." Paper presented at the International Conference on "Bioethical Issues of Intellectual Property in Biotechnology," Tokyo, Japan, 6–7 September 2004. http://www.ipgenethics.org/conference.htm.

MacGregor, Robert. "The Effects of Culture and Policy on Science in China." Unpublished paper, Rice University, May 4, 2004.

Maio, Giovanni. "The Embryo in Relationships: A French Debate on Stem Cell Research." *Journal of Medicine and Philosophy* 29, no. 5 (2004): 583–602.

Matthiessen-Guyader, Line. "Survey on Opinions from National Ethics Committees or Similar Bodies, Public Debate and National Legislation in Relation to Human Embryonic Stem Cell Research and Use." European Commission report, July 2004.

McLaren, Anne. "Ethical and Social Considerations of Stem Cell Research." *Nature* 414, no. 6859 (2001): 129–131.

Mieth, Dietmar. "Going to the Roots of the Stem Cell Debate: The Ethical Problems of Using Embryos for Research." *EMBO Reports* 1, no. 1 (2000): 4–6.

National Research Council and Institute of Medicine. *Guidelines for Human Embryonic Stem Cell Research* Washington, D.C.: National Academies Press, 2005.

National Review. "From the House, a Disgrace." June 20, 2005.

New Scientist. "Stem Cell Standoff," June 4, 2005, 6.

New York Times. "Faked Research on Stem Cells is Confirmed by Korean Panel." December 23, 2005, A8.

Nisbet, Matthew. "The Controversy over Stem Cell Research and Medical Cloning." Committee for the Scientific Investigation of Claims of the Paranormal, unpublished paper, April 2 2004.

———. "Public Opinion about Stem Cell Research and Human Cloning." *Public Opinion Quarterly* 68, no. 1 (2004): 131–54.

Normile, Dennis, and Charles Mann. "Asia Jockeys for Stem Cell Lead." *Science* 307, no. 5710 (2005): 660–64.

———. "Asian Countries Permit Research, With Safeguards." *Science* 307, no. 5710 (2005): 664.

Pichler, Franz. "Ethical Questions on the Funding of Human Embryonic Stem Cell Research in the Sixth Framework Programme of the European Commission for Research, Technological Development and Demonstration." *Innovation* 18, no. 2 (2005): 261–71.

Pollack, Andrew. "A Case of Stunted Growth: California's Stem Cell Program is Hobbled but Staying the Course." *New York Times,* December 10, 2005, B1.

President's Council on Bioethics. *Monitoring Stem Cell Research.* Washington, D.C.: President's Council on Bioethics, 2004.

Ray, Julie. "Medicine Meets Morality in Stem Cell Debate." *Gallup Poll,* August 5, 2003.

Reichardt, Tony. "Studies of Faith." *Nature* 432, no. 7018 (2004): 666–69.

Resnik, David. "Privatized Biomedical Research, Public Fears, and the Hazards of Government Regulation." *Health Care Analysis* 7 (1999): 273–87.

Rosenthal, Elisabeth. "Britain Embraces Embryonic Stem Cell Research." *New York Times,* August 24, 2004, 6.

Saad, Lydia. "Most Americans in the Dark about Stem Cell Research." *Gallup Poll,* July 20, 2001.

Seawright, Stephen. "A New Report says Mainland Stem Cell Research is Eclipsing the International Field, While Western Researchers Remain Bogged Down by Moral Concerns." *South China Morning Post,* February 24, 2005, 16.

Smith, Helen Lawton. "Regulating Science and Technology: The Case of the UK Biotechnology Industry." *Law & Policy* 27, no. 1 (2005): 189–212.

Stevens, Denise. "Embryonic Stem Cell Research." *Houston Journal of International Law* (Spring 2003).

Stolberg, Sheryl. "House Approves a Stem Cell Bill Fought by Bush." *New York Times,* May 2, 2005, A1.

Stone, Richard. "U.K. Backs Use of Embryos, Sets Vote." *Science* 289, no. 5483 (2000): 1269–70.

Tokyo Daily Yomiuri. "Japan Must Join Pursuit of Scientific Progress." May 23, 2005, 4.

United Press International (UPI). "German Chancellor Backs Stem Cell Research." June 14, 2005.

Wade, Nicholas, and Choe Sang-Hun. "Human Cloning was All Faked, Koreans Report." *New York Times,* January 10, 2006, A1.

Wade, Nicholas. "Korean Researchers to Help Others Clone Cells for Study." *New York Times,* October 19, 2005, A14.

———. "Korean Scientist Said to Admit Fabrication in a Cloning Study." *New York Times,* December 16, 2005, A1.

Walters, LeRoy. "Human Embryonic Stem Cell Research." *Kennedy Institute of Ethics Journal* 14, no. 1 (2004): 3–38.

Wang, Yanguang. "Chinese Ethical Views on Embryo Stem (ES) Cell Research." In *Bioethics in Asia in the 21st Century.* Edited by S. Y. Song, Y. M. Koo, and D. Macer, Tsukuba, Japan: Eubios Ethics Institute, 2003, 49–55.

Webb, Tim. "Laboratory to the World: The UK's Big Push on Stem Cells." *London Independent,* March 20, 2005, 15.

Chapter 7 Chimeras

Bulletin's Frontrunner. "Brownback Vows to Block Patents on Cloned Humans." May 21, 2002, 1.

Business Line. "Human Kidneys Grown in Mice." December 25, 2002.

Chapple, Christopher. *Nonviolence to Animals, Earth, and Self in Asian Traditions.* Albany, NY: State University of New York Press, 1993.

Clutton-Brock, Juliet. *Horse Power: A History of the Horse and the Donkey.* Cambridge, MA: Harvard University Press, 1992.

The Economist, "Xeno's Paradox." December 21, 1996.

Fukayama, Francis. "The Clone Traders: Advances in Human Biotechnology Have Highlighted the Need for International Regulation." *Financial Times,* May 18, 2002, 1.

Greene, Mark, et al. "Moral Issues of Human-Non-Human Primate Neural Grafting." *Science* 309, no. 5733 (2005): 385–86.

Henderson, Mark, and Adam Sage. "Scientists Create Chickens with Teeth." *Ottawa Citizen,* June 4, 2003, A11.

Heuser, Stephen. "GTC Gets Surprise Boost from EU." *Boston Globe,* June 3, 2006, F7.

Highfield, Roger. "Stem Cell Researchers Plan to Create Rabbit-Human Embryos." *London Daily Telegraph,* January 13, 2006, 10.

Janofsky, Michael. "11 Indicted in Cases of Environmental Sabotage." *New York Times,* January 21, 2006, A1.

Jiji Press Ticker Service. "Japan to Approve Creation of Animal-Human Embryo." November 6, 2001, 1.

Jim Lehrer News Hour Transcript. "Extended Interview: Eugene Redmond." July 2005.

———. "Extended Interview: Hank Greely." July 2005.

———. "Extended Interview: Irving Weissman." July 2005.

Johnson, Carolyn. "From Myth to Reality: Scientists Can Create Animals with the Cells of Other Species, but are these Chimeras Medical Marvels or High-Tech Monsters." *Boston Globe,* April 19, 2005, D1.

Johnston, Ian. "Embryos Key to Beating Genetic Killers." *The Scotsman,* January 13, 2006, 2.

Mott, Maryann. "Animal-Human Hybrids Spark Controversy." *National Geographic News,* January 25, 2005.

National Academies of Science. *Guidelines for Human Embryonic Stem Cell Research.* Washington, D.C.: National Academies Press, 2005.

New Zealand Press Association. "USA Company Engineering NZ Goats to Produce Blood Clotting Drug." January 13, 2006.

Normile, Dennis, and Charles Mann. "Asia Jockeys for Stem Cell Lead." *Science* 307, no. 5710 (2005): 660–64.

Normile, Dennis. "Japan: Human Cloning Ban Allows Some Research." *Science* 290, no. 5498 (2000): 1872.

Pharma Marketletter. "Moratorium in Xenotransplantation Urged." January 28, 1998.

Pittsburgh Post-Gazette. "Monsters on the Loose; Why Chimeras are so Very Much in the News these Days." May 9, 2005, B-7.

Radowitz, John von. "Scientists 'Create Hybrid Human-Rabbit Embryos.'" *Press Association,* August 14, 2003, 1.

Robert, Jason, and Francoise Baylis. "Crossing Species Boundaries." *American Journal of Bioethics* 3, no. 3 (2003): 1–13.

Ross, Emma. "Fertility Society Denounces Creation of Human Chimera Embryo." *Associated Press Worldstream,* July 2, 2003.

Shreeve, Jamie. "The Other Stem-Cell Debate." *New York Times,* April 10, 2005.

Tangley, Laura. "A Therapy for Cat Allergies, Thanks to Mice." *New York Times,* April 5, 2005, 6.

Thomas, Morgan. "Monsters of Our Minds." *Courier Mail,* July 9, 2003, 22.

The Times of India. "Clinical Trials of Pig Organ Transplant in 2 Years." February 7, 2000.

Weintraub, Arlene. "Crossing the Gene Barrier." *Business Week,* January 16, 2006, 72–80.

———. "What's Ethical and What Isn't?" *Business Week,* January 16, 2006, 76.

Weiss, Rick. "Cloning Yields Human-Rabbit Hybrid Embryo." *Washington Post,* August 14, 2003, A4.

———. "Human Brain Cells are Grown in Mice." *Washington Post,* December 13, 2005, A3.

Woolf, Marie. "Record Number of GM Animals being Bred in UK Laboratories." *London Independent,* December 4, 2005, 14.

Chapter 8 Pharmaceutical Companies and Biotechnology

Angell, Marcia. *The Truth About the Drug Companies.* New York: Random House, 2004.

Arnst, Catherine. "Biotech, Finally." *Business Week,* June 13, 2005.

Avorn, Jerry. *Powerful Medicines.* New York: Alfred A. Knopf, 2004.

Basu, Paroma. "International Patent Law—Boon or Bane of Biotech?" *Nature Biotechnology* 23 (2005): 13–15.

BIO. "Biotechnology Industry Organization Milestones 2004." undated report online at www.bio.org/speeches/pubs/milestone04/business.asp.

Caplan, Arthur. "Indicting Big Pharma." *American Scientist,* January–February 2005.

Chen, Zhiqi, and Alison McDermott. "International Comparisons of Biotechnology Policies." *Journal of Consumer Policy* 21 (1998): 527–50.

Cortina, Jose. "International Patent Laws Affect RTP Biotech Industry." Daniels, Daniels, and Verdonik, undated, http://www.d2vlaw.com/articles.htm.

Crosariol, Beppi. "Why Laws Really Love Biotech." *Toronto Globe and Mail,* June 7, 2006.

DiGrazio, Jeanne. "Patenting Human Genes." *Science and Technology Newsletter,* Bryn Mawr College, April 2002.

Drahos, Peter. "Developing Countries and International Intellectual Property Standard-Setting." *Journal of World Intellectual Property* 5, no. 5 (2002): 765–89.

Drinkard, Jim. "Drugmakers Go Furthest to Sway Congress." *USA Today,* April 26, 2005, 1B.

DrugResearcher.com. "Falling Innovation Levels Fuel Pharma/Biotech Collaborations." May 10, 2006, .

Eaton, Jonathan, Samuel Kortum, and Josh Lerner. "International Patenting and the European Patent Office: A Quantitative Assessment." *Patents, Innovation and Economic Performance, OECD Proceedings, 2004,* OECD, 2004, 27–52, http://www.oecd.org/dataoecd/12/4/33793381.pdf.

Friedman, Jane. "New Report Shows Biotech Companies Bring More New Drugs to Market than Big Pharmaceuticals." *DigitalJournal.Com,* April 21, 2006.

Gillis, Justin. "Biotech's Gains Again Outstrip Drug Giants." *Washington Post,* April 12, 2006, D1.

Gray, Nicole. "Sixth Annual Report of the World's Top 50 Pharma Companies: Untying the Gordian Knot." *Pharm Exec,* May 2005: 83–100.

Gwynne, Peter, and Guy Page. "Biotechnology: Personal Portraits of an Evolving Industry." *Science,* 287, no. 5461 (2000): 487.

Hamilton, Wanda. "Big Pharmaceuticals, Big Money." www.forces.org, August 24, 2001.

Hearn, Kelly. "Big Pharma's Deadly Experiments." *AlterNet,* June 9, 2006. Online at www.alternet.org.

Kassirer, Jerome. *On the Take.* New York: Oxford University Press, 2005.

Lai, Edwin. "International Intellectual Property Rights Protection and the Rate of Product Innovation." *Journal of Development Economics* 55 (1998): 133–53.

Levine, Daniel. "Big Pharmas Agreeing to Big Deals with Biotech Firms." *San Jose Business Journal,* June 20, 2003.

MacLeod, Calum. "China Makes Ultimate Punishment Mobile." *USA Today,* June 15, 2006, 8A.

Maurseth, Per Botolf, and Bart Verspagen. "Knowledge Spillovers in Europe: A Patent Citations Analysis." *Scandinavian Journal of Economics* 104, no. 4 (2002): 531–45.

Moynihan, Ray, and Alan Cassels. *Selling Sickness: How the World's Biggest Pharmaceutical Companies are Turning Us All into Patients.* New York: Nation Books, 2005.

Payne, January. "Still, A Cruel Diagnosis: In King's Death, a Lesson in Ovarian Cancer's Deadliness." *Washington Post,* February 14, 2006, F1.

Pollack, Andrew. "Signs That Biotech Has a Healthier Future." *New York Times,* April 4, 2006, C4.

———. "U.S. Finance Pulls Biotech Across Seas." *New York Times,* July 12, 2006, C1.

Romney, Lee. "Out-of-State Facility Demands Part of Stem Cell Research Royalties." *Los Angeles Times,* March 30, 2006, B3.

Sakakibara, Mariko, and Lee Branstetter. "Do Stronger Patents Induce More Innovation? Evidence from the 1988 Japanese Patent Law Reforms." NBER Working Paper Series, April 1999, unpublished paper.

Shah, Sonia. *The Body Hunters: How the Drug Industry Tests its Products on the World's Poorest Patients.* New York: New Press, 2006.

Suresh, N. "Global Biotech Crosses $60-Billion Mark." *BioSpectrum,* May 12, 2006.

Tremblay, Jean-Francois. "India Emerges in Drug Services." *Chemical and Engineering News* 83, no. 5 (2005): 15–17.

Vedantam, Shankar, and Marc Kaufman. "Doctors Influenced By Mention of Drug Ads." *Washington Post,* April 27, 2005, A1.

Vulliamy, Ed. "Medicine and Ethics: The Cell Struggle." *Guardian Newspaper Observer,* July 8, 2001, 18.

Wade, Nicholas. "Scientists Draft Rules on Ethics For Stem Cells." *New York Times,* April 27, 2005, 1.

Wall Street Reporter. "Successful New Models for Biotech Big Pharma Partnership." June 21, 2006.

Chapter 9 Ethics and Biotechnology

Bauer, Martin, John Durant, and Geoffrey Evans. "European Public Perceptions of Science." *International Journal of Public Opinion* 6, no. 2 (1994): 163–86.

Bonnicksen, A. *Crafting a Cloning Policy: From Dolly to Stem Cells.* Washington, D.C.: Georgetown University Press, 2002.

Caulfield, T., L. Knowles, and E. Meslin. "Law and Policy in the Era of Reproductive Genetics." *Journal of Medical Ethics* 30, (2004): 414–17.

Curtis, Michele. "Cloning and Stem Cells: Processes, Politics, and Policy." *Current Women's Health Reports* 3 (2003): 492–500.

Derthick, Martha, and Paul Quirk. *The Politics of Deregulation.* Washington, D.C.: Brookings Institution Press, 1985.

European Commission. "Europeans, Science & Technology." Special EuroBarometer 224. June 2005. (unpublished report).

Japan National Institute of Science and Technology Policy. "The 2001 Survey of Public Attitudes Toward and Understanding of Science & Technology in Japan." Unpublished January 2002 report.

Morgan, Derek. "Ethics, Economics and the Exotic: The Early Career of the HFEA." *Health Care Analysis* 12, no. 1 (2004): 7–26.

National Science Board. "Information Sources, Interest, and Perceived Knowledge." Unpublished 2004 report.

National Science Foundation. "Survey of Public Attitudes Toward and Understanding of Science and Technology." Unpublished 2001 report.

Nisbet, Matthew. "Public Opinion about Stem Cell Research and Human Cloning." *Public Opinion Quarterly* 68, no. 1 (2004): 131–54.

Quirk, Paul. *Industry Influence in Federal Regulatory Agencies.* Princeton, NJ: Princeton University Press, 1981.

Sagar, Ambuj, Arthur Daemmrich, and Mona Ashiva. "The Tragedy of the Commoners: Biotechnology and its Publics." *Nature* 18 (January 2000): 2–4.
Salter, Brian, and Mavis Jones. "Regulating Human Genetics: The Changing Politics of Biotechnology Goverenance in the European Union." *Health, Risk & Society* 4, no. 3 (2002): 325–40.
Science. "Europeans Apathetic Toward Science." December 10, 2001.
Smith, Helen Lawton. "Regulating Science and Technology: The Case of the UK Biotechnology Industry." *Law and Policy* 27, no. 1 (January 2005): 189–212.

Index

activism, 9, 25–26, 29, 31–32, 49, 59, 63, 64, 110–12
agribusiness, xii, 51, 67, 133, 137
Alzheimer's disease, 88, 104
animal-human hybrids, *See* chimeras
Argentina, 27, 49, 50, 67
Asia, xii, 6–7, 8–9, 40–43, 49, 51, 55, 60–67, 72, 74, 79–81, 85, 94–99, 109–12, 115, 132, 140
  assisted reproductive technology, 40–3
  biotech industry, 115
  chimeras, 109–12
  cloning, 72, 74, 79–81
  genetically modified food, 49, 51, 55, 60–68
  stem cell research, 85, 94–99, 109–12
assisted reproductive technology (ART), 24, 31–48
  types of, 32–34
  See fertility tourism; in vitro fertilization
Australia, 27, 42, 49, 66, 71
Austria, 37, 38, 56–58, 71, 86, 93, 99

Bacillus thuringiensis (Bt) toxin, 29–30, 54–55, 59, 61–65
Bangladesh, 60
Bayh-Dole Act, 6–7, 21
Belgium, 56–57, 71
Big Pharma, xii, 1–2, 115–19, 121–23, 126
  alliances
  See biotech-pharmaceutical company
biotech globalization (new globalization), ix, xi, xiii, 12–13, 131, 143
biotech industry, 78, 92, 115–119, 123–9, 134
  See biotech-pharmaceutical company
  alliances; licensing agreements; partnerships; patents; venture capital
biotech-pharmaceutical company
  alliances, 116–19
Brazil, 49–50, 56
Brown, Louise, x, 33–34, 37
Bt-crops *See Bacillus thuringiensis*
Buddhism, 41–42, 94
Bulgaria, 139
Bush, George W., 17, 36, 73, 89–91

Canada, 4, 20, 26, 49–50, 56, 67, 71, 107, 121–22
  chimeras, 26, 107
  genetically modified food, 49–50, 56
  research and development, 4, 20
Cartagena Protocol on Biosafety, 63
Catholicism, 31–32, 34, 36, 40, 44–48, 71–72, 94, 105, 110, 136
  assisted reproductive technology, 31–32, 34, 36, 40, 44–48
  chimeras, 105, 110
  cloning, 71–72
  chimeras, ix, x, xii, 1, 12, 22, 25–26, 80, 101–14
  See xenotransplantation

China, ix, x, 1, 3, 9, 12, 16, 22, 26, 28, 29, 42–43, 49–51, 56, 60–64, 67, 69–71, 79–81, 85–86, 95, 96–99, 103, 107, 109–12, 126, 128–29, 132, 142
  assisted reproductive technology, 42–43
  chimeras, 26, 80, 103, 107, 109–12
  cloning, 69–71, 79—81
  cotton, 61–62
  genetically modified food, 1, 29, 49–51, 56, 61–64, 67
  human clinical trials, 128, 142
  human organ trade, 128–29
  labeling of, 63, 64
  medical tourism, 128
  patents, 126
  rice, 29, 51, 61, 63–64
  stem cell research, 85–86, 95, 96–99, 111
Christianity, 87, 102, 109–10, 136
  See Catholicism
Clinton, William Jefferson, 17, 19, 89
cloning, ix, x, xii, 12, 24, 52, 69–83, 97, 110, 135–36, 140
  animal, 69–70, 72, 75–76, 79–80
  deficiencies of, 72
  definitions of, 70, 73
  economic payoffs, 69, 82–83
  reproductive, 69–74, 76, 77, 78, 81–2
  therapeutic (nonreproductive), 69–82, 110
  See Dolly, the sheep
contraband technology, xi, 1, 28–30
contraband trade, 51, 59, 63, 68, 134
"country-shopping," x–xi, 1, 10, 25–28, 33, 48, 51, 68, 70, 96, 99–100, 112, 113–14, 120–21, 122–23, 126, 131, 134, 137, 142

decision making, ix, x, 24, 30, 31, 48, 134–43
Denmark, 37, 38, 56–7, 71
deoxyribonucleic acid (DNA), 78, 110–11, 113, 139
deregulation, xi, 2, 16–18, 131–32
diabetes, 26, 88, 127
Dolly, the sheep, x, 1, 15, 27, 69, 72, 75, 76, 81–82, 108, 141

economic globalization (old globalization), ix, xi, 11–13, 131, 143
embedded autonomy, 9, 10
ethics, ix, xii, xiii, 12, 19, 23, 24, 26, 28, 30, 31–32, 33, 35–36, 69, 71–72, 76, 85–88, 101–7, 109–11, 113–14, 120, 122–23, 132–43
  Big Pharma, 121–23
  chimeras, 26, 101–7, 109–11, 113–14
  cloning, 69, 71–72, 76, 135–36
  in vitro fertilization, 31–32, 33, 35–36
  stem cell research, 85–88, 135–36
  See decision making; international political economy (IPE); public education; religious political economy (RPE)
European Commission, 57–58, 60, 66, 94, 138, 141
European Union (EU), xii, 2, 4, 6–7, 10, 24, 25, 34, 37–40, 43–48, 49, 51, 54–60, 63, 65, 66–67, 72, 75–79, 82, 91–94, 99, 107–9, 115–16, 124, 138–41
  assisted reproductive technology, 37–40, 43–48
  biotech industry, 78, 115–16
  chimeras, 107–9
  cloning, 72, 75–79
  genetically modified food, 49, 54–60, 63, 65, 66–68, 138
  labeling of, 54–55, 57–59, 63, 65, 66, 138
  restrictive stance on, 56–60
  patents, 2, 124, 128
  research and development, 4
  stem cell research, 91–94, 99
experts, 1, 7–8, 10, 19, 23–24, 27, 30, 37, 79, 85, 119–20, 131, 135, 142

fertility tourism, 45–48
Finland, 37, 71
Flavr Savr tomato, 53–54
France, ix, 3, 4, 5, 7, 15, 20, 37–40, 45, 56–58, 67, 69, 71, 72, 75, 77–79, 81, 93–94, 99, 108, 121, 132–33
  assisted reproductive technology, 37–40, 45
  biotech industry, 78
  chimeras, 108
  cloning, 69, 72, 75, 77–79, 81
  genetically modified food, 56–58, 67
  research and development, 4, 15, 20
  stem cell research, 93–94, 99

genetically modified food, ix, x, xi–xii, 12, 24–25, 28–30, 49–68, 85, 134, 137–38, 141
  industry concentration, 137
  labeling of, 49–51, 54–59, 63, 64–68, 138
  safety of, 49, 51–54, 59–60, 63, 67, 141
  See Bacillus thuringiensis (Bt) toxin
Germany, ix, 3, 4, 5, 7, 10, 20, 22, 23, 31–32, 38, 39–40, 43–44, 47–48, 56–58, 67, 69, 71, 72, 75, 77–79, 81, 86, 93, 99, 107, 108–9, 125, 132–33
  assisted reproductive technology, 31–32, 38, 39–40, 43, 47–48
  biotech industry, 78
  chimeras, 107, 108–9
  cloning, 69, 72, 75, 77–79, 81
  genetically modified food, 56–58, 67
  history of, and regulation, 10, 31–32, 38, 47–48, 77, 81
  patents, 125
  research and development, 4, 20
  stem cell research, 86, 93, 99
globalization, ix–xi, xiii, 3–4, 10, 11–13, 18, 19, 25–30, 33, 82–83, 96–97
  See biotech globalization; "country shopping"; economic globalization;knowledge mobility; "scientist-buying"
Great Britain, ix, 1, 3, 4, 7, 10, 15–16, 18, 20, 22, 23, 27, 32–34, 38–40, 45, 46, 58–60, 69–70, 72, 74, 75–78, 80–81, 86, 91–93, 96, 99, 107–8, 110–11, 132, 136
  assisted reproductive technology, 38–40, 45, 46
  biotech industry, 78, 92
  chimeras, 107–8, 110–11
  cloning, 69–70, 72, 74, 75–77, 80–81, 99
  genetically modified foods, 58–60
  regulation in, 10, 16, 18, 38–39, 59–60, 136–37
  research and development, 4, 20
  stem cell research, 86, 91–93, 96, 99
Greece, 37–38, 45, 56–57, 71

health care, 126–29
higher education, ix, xi, 1, 2, 4, 5–10, 15–16, 18–23, 131, 134, 140
  commercialization of, 5–7, 19
  and corporate partnerships, xi, 5, 15–16, 18, 20–22, 131
  and patents, 20–22
  See also professors; experts
Hinduism, 41, 110
Hong Kong, 43, 69, 79
human clinical trials, 128–29, 142
human embryonic stem cells (hES), 23, 76, 85–99, 103, 106–8, 113–14, 119–20
Human Fertilisation and Embryonic Authority (HFEA), 107–8
Hungary, 37–38
Hwang, Woo Suk, 16, 95–96, 141

Iceland, 71
India, ix, x, 12, 22, 26, 41–43, 49, 60–61, 64–67, 71, 80–81, 95, 98, 109, 111–12, 125–26, 128, 132, 142
  assisted reproductive technology, 41–43
  chimeras, 26, 109, 111–12
  cloning, 80–81

India (*continued*)
genetically modified food, 49, 60–61, 64–67
labeling of, 65
human clinical trials, 128, 142
medical tourism, 112, 128
patents, 125–26, 128
stem cell research, 95, 98
Indonesia, 60–61
intellectual property, xii–xiii, 6, 116, 120, 123–26
See patents
international political economy (IPE), xiii, 132–33
Internet, 3–4, 7–8, 12, 13, 24, 28, 141
in vitro fertilization, ix, x, xi, 1, 12, 24, 31–39, 41–48, 49, 71, 76, 85, 88, 89, 91–92, 97
definition of, 32–34
strict regulation of, 43–48
Ireland, 31, 34, 43–44, 46–47, 48, 93, 133
Islam, 72, 87, 136
Israel, 37, 71
Italy, 7, 31–32, 43–45, 47, 48, 56–57, 60, 71, 75, 93, 121, 133
assisted reproductive technology, 43–45
cloning, 71, 75
genetically modified food, 56–57, 60

Japan, ix, 3, 4, 5, 20, 22, 26, 27, 29–30, 41–42, 60, 65–67, 81, 98–99, 109, 112, 125, 126, 132–33, 140
assisted reproductive technology, 41–42
chimeras, 26, 109, 112
cloning, 81
genetically modified food, 29–30, 65–66
labeling of, 65–66
patents, 125, 126
public education, 140
research and development, 4, 20
stem cell research, 98–99
Judaism, 87, 136

knowledge mobility, 7–8, 25–30, 70, 82–83, 96–97, 99, 114, 120–21, 134, 137–38, 141–43
Korea, ix, x, 5, 20, 22, 71, 80, 81, 95–96, 111
stem cell research, 95–96, 111
See also South Korea

licensing agreements, 116, 117, 118–19
lobbying, 36–37, 49, 51, 52, 55, 67–68, 122, 133
Luxembourg, 56–57

Malaysia, 61, 69–70, 79
market capitalism, 16–20, 78
medical tourism, 27, 33, 112, 128–29, 131
Mexico, 49, 71, 128–29
Monsanto, 51, 54, 61, 62, 64

Netherlands, 4, 7, 38, 46, 58, 71, 72, 75
New Zealand, 41–42, 66, 71
nongovernmental organizations (NGOs), ix, 1–2, 9, 10, 12, 16–17, 19, 23, 59, 67
nonprofit sector, ix, 4, 8, 10, 23
Norway, 38, 75

Pakistan, 60
Paraguay, 49
Parkinson's disease, 26, 88, 112–13
partnerships, 117–19
patents, xi, xii–xiii, 2, 6–7, 15–16, 19, 20–22,106, 116, 120, 123–26
pharmaceutical companies, global, *See* Big Pharma
Philippines, 49, 61, 79
Poland, 37–38, 93
Portugal, 58, 79
President's Council on Bioethics, 36, 74, 88
private sector, ix, xi, 3, 5, 6, 7–10, 18, 19, 22, 62, 64, 65, 89–90, 100, 105, 114
professors, 6–7, 15–16, 21–22
Protestantism, 40, 94, 105, 136

public education, 139–41
public opinion, x, xii, 24, 35–36, 38, 40–42, 48, 51, 55–56, 59–60, 66, 67, 74–75, 77–82, 85, 91–95, 110–11, 116, 121, 131–33, 136–37, 139–41
  assisted reproductive technology, 35–36, 38, 40–42, 48
  Big Pharma, 116, 121
  chimeras, 110–11
  cloning, 74–75, 77–82
  genetically modified food, 51, 55, 59–60, 66, 67
  stem cell research, 85, 91–95
public sector, ix, xi, 1, 2, 9, 10, 16–20, 23, 62, 64, 100, 107, 111, 114, 119–20
  industry standards, 100
  reshaping of, 16–20
  See deregulation

Reagan, Ronald, 16–18
recombinant DNA (deoxyribonucleic acid), 50–52
Redmond, Eugene, 26, 112–13
religion, x, xii, xiii, 12, 24, 28, 31–32, 40–42, 44–48, 71–74, 76, 79, 80, 82, 85, 87–88, 91–92, 94–95, 105, 108, 109–11, 131–36, 139, 142
  assisted reproductive technology, 40–42, 44
  chimeras, 105, 108, 109–11
  cloning, 71–74, 76, 79, 80, 82, 135–36
  Eastern versus Western, 40–42, 71–72, 79, 94, 109–11, 136
  fundamentalism, 12, 71–72, 142
  and life, beginning of, 71–72, 135–36
  the "religious right," 36–37
  stem cell research, 85, 87–88, 91–92, 94–95, 135–36
  versus science, 135, 139
  See Catholicism; Christianity; Islam;
religious political economy (RPE)
religious political economy (RPE), xiii, 133
research and development (R&D), 3–5, 7, 10, 15–22, 119–20
Romania, 49
Russia, 66

science parks, 22
scientific publications, xi, 3, 15, 28, 114
"scientist-buying," xi, 27, 51, 96–97, 99, 131
Singapore, 26–27, 69–70, 79, 85–86, 95–96, 99
  stem cell research, 85–86, 95–96, 99
Slovenia, 46
South Africa, 49
South Korea, 4, 5, 65, 72, 77, 85–86, 95, 99, 132, 141
  stem cell research, 85–86, 95, 99
space program, the, 3–4
Spain, 26, 37, 38, 45, 49, 58, 71, 72, 79, 93–94, 96
StarLink corn "escape," 54–55, 59
start-up companies, 6–7, 15–16, 21–22
state, the x–xi, 1–2, 7–10, 12, 15, 16, 18, 26, 37, 120, 131, 132, 134, 141–42
stem cell research, ix, x, xii, 1, 12, 16, 22, 24–26, 80, 85–100, 119–20, 128, 135–36, 141–42
  See chimeras
St. Kitts Biomedical Foundation, 26, 112–13
St. Kitts island, Caribbean, 1, 25–26, 112–13
"substantial equivalence," 53–54, 67
Sweden, 4, 7, 22, 38, 71, 75, 139
Switzerland, 29–30, 38, 60, 71, 92–93
Syngenta, 29–30, 51, 59, 62, 65

Taiwan, 95
Thailand, 41–42, 61, 66, 128
Thatcher, Margaret, 16, 18
"transformation," 50
Turkey, 59, 139

United Nations (UN), 72–73, 77, 138
United States (U.S.), ix, xii, 1, 2, 3, 4–5, 6, 10, 15–18, 20–23, 25, 26–27, 32–37, 40, 48, 49–68, 71, 72, 73–75, 78, 86, 88–91, 96–99, 103, 105–7, 113, 115–16, 121–22, 124, 128, 132–33, 139–41
  assisted reproductive technology, 34–37, 40, 48, 49, 88
  biotech industry, 78, 115–16
  chimeras, 26, 103, 105–7, 113
  cloning, 1, 71, 72, 73–75, 88
  drug costs, 121–22
  genetically modified foods, 49–68, 88, 139
    agribusiness lobbying, 49, 51, 52, 55, 67–68, 133
    exporting, 49, 55, 57–58, 63, 67
    labeling of, 51, 54, 55–56, 65, 66–67
    See StarLink corn “escape”
  patents, 2, 124, 128
  regulation, 10, 16–18, 27, 34–37, 128
  research and development, 4–5, 15–16, 18, 20–22
  stem cell research, 26, 86, 88–91, 96–99
U.S. Congress, 6, 17, 34, 51, 55, 68, 73–74, 86, 89, 90, 106–7, 122
U.S. Department of Agriculture, 52, 54
U.S. Environmental Protection Agency (EPA), 54
U.S. Food and Drug Administration (FDA), 52–54, 56, 58, 75, 117, 119, 121, 128
universities, *See* higher education
Uruguay, 49

venture capital, 117, 119

Warnock Report (1984), 23
Wilmut, Ian, 69, 72, 108
World Trade Organization, 57–58, 125–26

xenotransplantation, 69, 108, 111